Rudolf Höfling

Die Flugzeuge und Hubschrauber des Österreichischen Bundesheeres

Airplanes and Helicopters of the Austrian Armed Forces

Rudolf Höfling

Die Flugzeuge und Hubschrauber des Österreichischen Bundesheeres
Von 1955 bis heute

Airplanes and Helicopters of the Austrian Armed Forces
From 1955 to Today

W-H EDITION WINKLER-HERMADEN

Auf dem Umschlagbild sind folgende Fluggeräte zu sehen:
Saab J29F Tonne (oben links), Saab 35OE Draken (oben rechts), Bell AB 204B (unten links), Jakowlew Jak-18 (unten rechts).

Cover photos clockwise from top left:
Saab J29F Tunnan, 35OE Draken, Bell AB 204B und Yakovlev Yak-18.

Übersetzung / Translation: Wolfgang Geißler und Wolfgang M. Buchta

www.edition-wh.at
Printed in the EU
ISBN 978-3-9504625-7-9

Inhalt / Contents

De Havilland Vampire F.B.5 der 72. Squadron der Royal Air Force aus North Weald zu Besuch bei der RAF Station Schwechat, Anfang der 1950er-Jahre.

De Havilland Vampire F.B.5 of the 72nd Squadron of the Royal Air Force from North Weald visiting the RAF station Schwechat, early 1950s.

Einleitung

Mit der Unterzeichnung des Österreichischen Staatsvertrages am 15. Mai 1955, mit dem die Souveränität Österreichs wiederhergestellt und der Abzug der alliierten Besatzungstruppen bis zum Ende desselben Jahres fixiert wurde, begann der unmittelbare Aufbau der neuen österreichischen Luftstreitkräfte. So wurde am 15. Juli 1955 im Bundeskanzleramt das Amt für Landesverteidigung etabliert, und am folgenden 13. September entstand unter der Leitung von Major Josef Bizek, einem Piloten, der zuvor bei der sogenannten B-Gendarmerie (eine Tarnbezeichnung, da von den alliierten Besatzungsmächten ein offizielles Heer noch nicht erlaubt war) gedient hatte, die untergeordnete Abteilung III/L (Luft), welche damals über nur zwei Mitarbeiter und ein bescheidenes Büro in der Dominikanerbastei im 1. Wiener Gemeindebezirk verfügte.

Der Aufbau der österreichischen Luftstreitkräfte war erheblich schwieriger als jener des österreichischen Heeres, welches auf das Personal und die Ausrüstung der B-Gendarmerie zurückgreifen konnte, die bereits vor der Unterzeichnung des Österreichischen Staatsvertrages gegründet und von den Vereinigten Staaten von Amerika mit Ausrüstungsgegenständen und Waffen für etwa 6000 Mann ausgerüstet worden war.

Zur tatsächlichen Geburtsstunde der österreichischen Luftstreitkräfte wurde der 9. Dezember 1955. Als an diesem Tag der Polizei-Oberstleutnant Gustav Hauck mit der vor Ort zusammengebauten Jakowlew Jak-18 (3A-AB) – einem Geschenk der Sowjetunion – um 10:01 Uhr erstmals vom Fliegerhorst Brumowski im niederösterreichischen Langenlebarn startete, war dies nach mehr als 17,5 Jahren Pause der erste Flug eines Flugzeuges mit österreichischen Markierungen. Mit diesen Jak-18-Schulflugzeugen wurde kurz danach die Schulung kriegserfahrener Piloten und des Bodenpersonals aufgenommen.

Die Erstausrüstung des Österreichischen Bundesheeres war zu Lande und in der Luft durchaus als bescheiden zu bezeichnen und bestand hauptsächlich aus Geschenken und sehr günstig erworbenem, gebrauchtem Gerät der ehemaligen Besatzungsmächte. So verfügten die österreichischen Luftstreitkräfte Anfang des Jahres 1956 nur über je vier Jakowlew Jak-11 und Jak-18 zur Grundschulung – Geschenke der UdSSR – sowie über einen in Frankreich erworbenen Hubschrauber des Typs Bell 47G2. Daher war es die vordringlichste Aufgabe, modernes fliegendes Gerät zu beschaffen. In diesen Monaten trafen im Bundesministerium für Landesverteidigung von verschiedenen Herstellern und Zwischenhändlern Angebote über einige moderne Flugzeugtypen ein, welche von der Dassault Mystere IV über die De Havilland Vampire und die Hispano Aviacion HA-200 Saeta bis zur North American F-86 Sabre reichten.

Auf den folgenden Seiten sind alle vom Österreichischen Bundesheer beschafften und geleasten Flugzeug- und Hubschraubertypen von 1955 bis zur Gegenwart chronologisch angeführt.

Introduction

With the signing of the Austrian State Treaty on 15 May 1955, which restored the sovereignty of Austria and fixed the withdrawal of the Allied occupying forces until the end of the same year, the immediate establishment of the new Austrian Air Force began. Thus, on 15 July 1955, the Office of National Defence was established in the Federal Chancellery, and on the following 13 September under the direction of Major Josef Bizek, a pilot, who previously had served in the so-called B-Gendarmerie (a camouflage, since by the Allied occupying powers an official army was not yet allowed) the subordinate department III / L (air), which then had only two employees and a modest office in the Dominican Bastion in the 1st district of Vienna was created.

The construction of the Austrian Air Force was considerably more difficult than that of the Austrian Army, which could draw on the staff and equipment of the B-Gendarmerie, which had been founded before the signing of the Austrian State Treaty and had been equipped by the United States of America with equipment and weapons for about 6000 men. The actual birth of the Austrian Air Force was the 9 December 1955. On that day, the police lieutenant colonel Gustav Hauck started at 10:01 am for the first time from Airbase Brumowski in Langenlebarn in Lower Austria with the locally assembled Yakovlev Yak-18 (3A-AB) – a gift from the Soviet Union –, this being after a more than 17.5 years break the first flight of an airplane with Austrian markings. Shortly after that, the training of experienced pilots and ground personnel started with these Yak-18 training planes.

The original equipment of the Austrian Armed Forces was on land and in the air relatively modest and consisted mainly of gifts and very cheaply acquired, used equipment of the former occupying powers. Thus, at the beginning of 1956, the Austrian Air Force had only four Yakovlev Yak-11 and Yak-18 for basic training – gifts from the USSR – and one Bell 47G2 helicopter acquired in France. Therefore, the most urgent task was to procure modern flying equipment. During these months, the Federal Ministry of Defence received offers from various manufacturers and distributors for some modern aircraft types, ranging from the Dassault Mystere IV to the De Havilland Vampire and the Hispano Aviacion HA-200 Saeta to the North American F-86 Sabre.

On the following pages, all types of airplanes and helicopters bought and leased by the Austrian Armed Forces from 1955 to the present day are listed chronologically.

Mitte der 1970er-Jahre entschied sich das Österreichische Bundesheer, ein effizientes und modernes Luftraumüberwachungssystem, das als „Goldhaube" bezeichnet wurde, zu etablieren. Dieses System besteht aus drei ortsfesten Radarstationen in Kärnten, Niederösterreich und Salzburg sowie ergänzenden Einrichtungen, die eine Überwachung des österreichischen Luftraums 365 Tage im Jahr und 24 Stunden pro Tag ermöglichen.

In the mid-1970s, the Austrian Armed Forces decided to establish an efficient and modern airspace monitoring system, which was called the "golden cap". This system consists of three stationary radar stations in Carinthia, Lower Austria and Salzburg as well as additional facilities that allow monitoring of Austrian airspace 365 days a year, 24 hours a day.

Jakowlew Jak-18

Das sowjetische Schulflugzeug Jakowlew Jak-18 (NATO-Code „Mouse“) flog erstmals im Jahr 1945. Vier Exemplare dieses unbewaffneten Typs …

The Soviet training airplane Yakovlev Yak-18 (NATO-Code “Mouse”) made its maiden flight in 1945 and four copies of this unarmed type …

… trafen gemeinsam mit vier weiteren Jakowlew Jak-11-Schulflugzeugen als Geschenk der Sowjetunion im November 1955 per Bahntransport in Österreich ein.

… arrived together with four training aircraft Yakovlev Yak-11 – all presents of the Soviet Union – in Austria by rail transport in November 1955.

Ein russischer und ein österreichischer Dolmetscher unterstützten die Techniker des Österreichischen Bundesheeres beim Zusammenbau der Jakowlew-Schulflugzeuge.

A Russian and an Austrian translator supported the technicians of the Austrian Armed Forces during the assembly of the Yakovlev Yak-11 trainer.

Jakowlew Jak-11

Die vier ehemals sowjetischen Jakowlew Jak-11 (NATO-Code „Moose“) auf diesem Foto wurden 1958 in Langenlebarn aufgenommen.

These four former Soviet Yakovlev Yak-11 (NATO-Code “Moose”) were photographed in Langenlebarn in 1958.

Als Fortgeschrittenentrainer konnte die Jakowlew Jak-11 mit einem 12,7 mm Maschinengewehr UBS (Uniwersalnji Beresin Synkhronnom) und zwei 50-Kilo-Bomben ausgerüstet werden.

As an advanced trainer the Yakovlev Yak-11 could be equipped with a 12.7 mm machine gun UBS (Uniwersalnji Beresin Synkhronnom) and two 50 kilo bombs.

Im Jänner 1956 führten Oberstleutnant Gustav Hauck (Dritter von rechts) und Josef Lechner (mit Fallschirm) in Langenlebarn ihren ersten Flug mit einer Jakowlew Jak-11 durch.

In January 1956 Lt Col Gustav Hauck (3rd from right) and Josef Lechner (with parachute) made their first flight with a Yakovlev Yak-11 in Langenlebarn.

Als während der Libanon-Krise Transportflugzeuge der United States Air Forces in Europe ab dem 16. Juli 1958 damit begannen, ohne viel Aufhebens die kürzeste Flugroute von Süddeutschland nach Italien zu nehmen – die allerdings über Tirol führte –, wurden zur Unterstreichung der diplomatischen Proteste der österreichischen Regierung gegen diese Luftraumverletzungen neben drei De Havilland D.H.115 Vampire auch zwei Jakowlew Jak-11 für Patrouillenflüge auf den Flughafen Innsbruck-Kranebitten verlegt. Als die Flugzeuge in Kranebitten ihre Flüge aufnahmen, war die Verlegung der US-Truppen nach Italien allerdings bereits beendet.

During the Lebanon crisis, as of 16 July 1958, Unites States Air Force in Europe transport planes began, just like that, to take the shortest route from southern Germany to Italy – which, however, crossed Tyrol. As a diplomatic protest against these airspace violations the Austrian government relocated in addition to three De Havilland DH115 Vampires also two Yakovlev Yak-11 for patrol flights to the airport Innsbruck-Kranebitten. However, when the planes began their flights in Kranebitten, the transfer of US troops to Italy had already ended.

Bell 47G2 & Agusta-Bell 47G2

Ende Jänner 1956 reiste eine dreiköpfige Delegation des Österreichischen Bundesheeres nach Paris und übernahm dort bei der Züricher Firma Forrer-Fenwick AG einen in den USA bei der Bell Helicopter Company gebauten Bell 47G2, der damit zum ersten Militärhubschrauber Österreichs wurde.

At the end of January 1956, a three-member delegation of the Austrian Armed Forces travelled to Paris, where it received a Bell 47G2 built by the Bell Helicopter Company in Zurich at Forrer-Fenwick AG, becoming the first military helicopter in Austria.

Cessna 172

Für Schulungszwecke erwarb das Österreichische Bundesheer im Jänner 1957 eine Cessna 172, die allerdings nur bis zum Ende des folgenden Jahres im Dienst verblieb.

For trainings purposes the Austrian Armed Forces obtained a Cessna 172 in January 1957, which stayed in service only until the following year.

Cessna 182

Gleichzeitig mit der Cessna 172 stellte das Österreichische Bundesheer im Jänner 1957 eine ebenfalls viersitzige Cessna 182 für Schulungszwecke in Dienst, der im Jahr 1959 eine zweite Maschine dieses Typs folgte.

In January 1957, at the same time as the Cessna 172, the Austrian Armed Forces commissioned an equally four-seater Cessna 182 for training purposes, followed by a second airplane of this type in 1959.

Zlin Z-126

Für die Fortgeschrittenen- und die Kunstflugschulung wurden im Jänner 1957 in der ČSR vier Zlin Z-126 erworben.

For advanced and aerobatic training, four Zlin Z-126s were acquired in ČSR in January 1957.

Die bei der Firma Moravan-Otrokovice, im 60 Kilometer östlich von Brünn gelegenen Otrokovice gebauten Z-126 blieben bis zum Juni 1965 im Dienst der österreichischen Luftstreitkräfte.

The Z-126, built by Moravan-Otrokovice, Otrokovice, 60 kilometres east of Brno, remained in service with the Austrian Air Force until June 1965.

Die Zlin Z-126 mit der Werksnummer 809 und der Registrierung 2A-AL steht hier im Dienst bei der Fluggruppe 1.

The Zlin Z-126 with the serial number 809 and the registration 2A-AL was in service with Fluggruppe 1 (Flying Group 1).

Piper PA-18 Cub

Ab dem März 1957 wurden vom Österreichischen Bundesheer acht Piper PA-18/95 und zwei PA-18/150 zur Schulung beschafft.

From March 1957 on the Austrian Armed Forces obtained eight Piper PA-18/95 and two PA-18/150 for training purpose.

Die Piper PA-18/95 wurden von 91 PS leistenden Continental C-90-Triebwerken und die PA-18/150 von 152 PS leistenden Lycoming O-320-Motoren angetrieben.

The Piper PA-18/95 were powered by a Continental C-90 engine of 91 h.p. and the P-A-18/150 by a Lycoming O-320 engine of 152 h.p.

Diese Piper PA-18/95, hier aufgenommen am 31. August 2002, ein ehemaliges Flugzeug des Österreichischen Bundesheeres, diente mit ziviler Registrierung und neuem Propeller in Langenlebarn als Schleppflugzeug für Segelflugzeuge.

This Piper PA-18/95, here photographed on 31 August 2002, a former airplane of the Austrian Armed Forces served with civil registration and new propeller in Langenlebarn as a towplane for gliders.

De Havilland D.H.115 Vampire T.11/T.55

Am frühen Nachmittag des 26. März 1957 trafen die ersten drei De Havilland D.H.115 Vampires, eine T.11 und zwei T.55, geflogen von zwei Bundesheerpiloten und einem englischen Fluglehrer, in Wien-Schwechat ein.

In the early afternoon of 26 March 1957, the first three De Havilland D.H.115 Vampires, one T.11 and two T.55, arrived in Vienna-Schwechat, flown by two pilots of the Austrian Armed Forces and an English flight instructor.

Am 13. Jänner 1961 erhielt die Jabo-Schulstaffel (Jagdbomber-Schulstaffel) in Graz-Thalerhof zwei weitere De Havilland D.H.115 Vampire T.55, und am 8. Dezember 1964 starteten schließlich die letzten drei Vampire T.11 vom De Havilland-Werksflugplatz Hatfield nach Linz-Hörsching.

On 13 January 1961, the Fighter Bomber School Squadron in Graz-Thalerhof received two more De Havilland DH115 Vampires T.55, and on 8 December 1964, finally launched the last three Vampires T.11 from the De Havilland Factory airfield Hatfield to Linz-Hörsching.

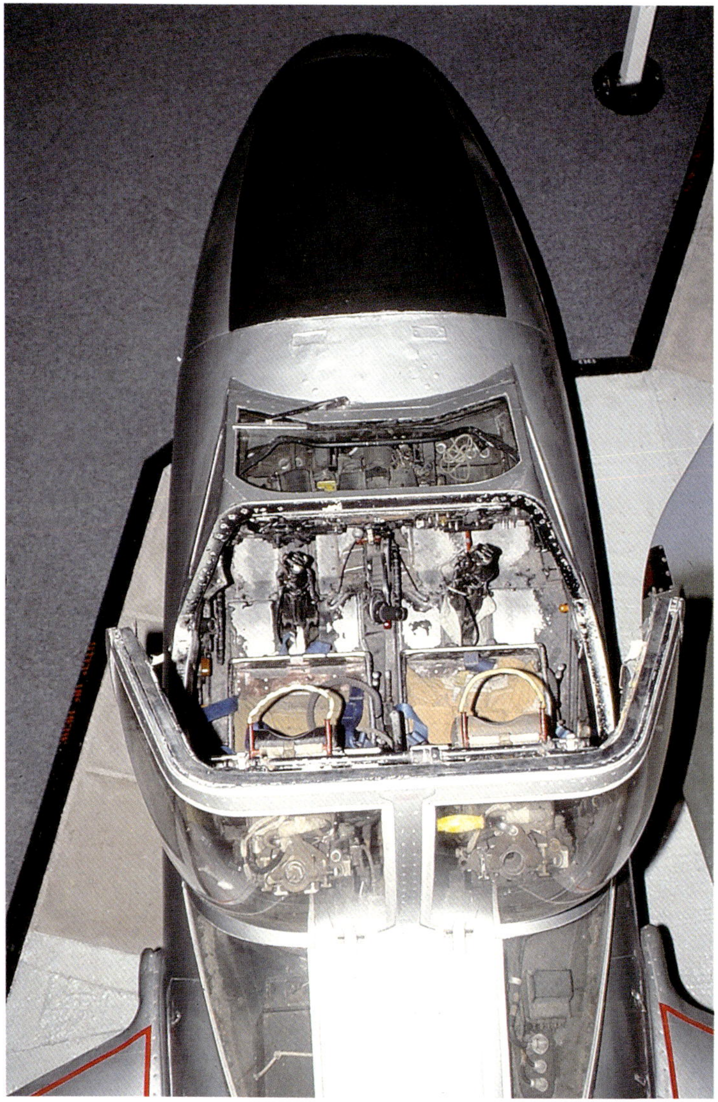

Die De Havilland D.H.115 T.11 Vampire unterscheidet sich von der Exportversion T.55 nur durch eine geänderte Instrumentierung.

The De Havilland D.H.115 T.11 Vampire differs from the export version T.55 only by different instruments.

Fiat G.46

Im April 1957 bestellte das Verteidigungsministerium fünf Fiat G.46-Schulflugzeuge, die im folgenden Jahr der Fliegerschulkompanie 2 in Linz-Hörsching zugeteilt wurden.

In April 1957 the Austrian Ministry of Defence ordered five Fiat G.46 training airplanes which were assigned to the Fliegerschulkompanie 2 (Pilots Training Squadron 2) in Linz-Hörsching in the following year.

Anfang der 1960er-Jahre zeigte es sich, dass viele der Schulflugzeuge des Österreichischen Bundesheeres „abgeflogen" waren. Zusätzlich wurde die Ersatzteilbeschaffung für die zahlreichen Flugzeugtypen zunehmend schwieriger, was zu einer Typenbereinigung mit der Saab 91 Safir führte, weshalb im Mai 1963 auch alle Fiat G.46 außer Dienst gestellt wurden.

In the early 1960s, it became apparent that many of the Austrian Air Force training airplanes had become obsolete. In addition, the replacement parts procurement for the numerous airplane types became increasingly difficult, which led to a type adjustment with the Saab 91 Safir, which is why in May 1963 all Fiat G.46 were decommissioned.

Die einzige „überlebende“ Fiat G.46 des Österreichischen Bundesheeres mit der Registrierung „3A-BB“ wird in Zeltweg aufbewahrt. (Diese Aufnahme stammt vom 9. Juni 2005.)

The only surviving Fiat G.46 of the Austrian Armed Forces with the registration “3A-BB” is preserved in Zeltweg. (This picture was taken 9 June 2005.)

Saab 17A

Die einzige Saab 17A, welche die österreichischen Luftstreitkräfte am 23. August 1957 als Zieldarsteller kauften, hatte davor bei der Svenska Flygvapnet (Schwedische Luftwaffe) als Sturzkampfbomber gedient.

The only Saab 17A which the Austrian Air Force purchased on 23 August 1957 for target towing had been used before as dive bomber of the Svenska Flygvapnet (Swedish Air Force).

1954 wurde die Saab 17A bei der Schwedischen Luftwaffe ausgeschieden und von der Svensk Flygtjänst AB zu einem Zielschleppflugzeug umgebaut.

In 1954 this Saab 17A was retired from the Svensky Flygvapnet and was modified as a target towing airplane by the Svensk Flygtjänst AB.

Fairey Firefly TT.1

Aufgrund regelmäßig auftretender technischer Probleme mit der Saab 17A wurden vom Österreichischen Bundesheer ab 1963 bei der Svensk Flygtjänst AB in Stockholm Fairey Firefly TT.1 …

Due to regularly occurring technical problems with the Saab 17A the Austrian Armed Forces leased from 1963 from the Svensk Flygtjänst AB in Stockholm Fairey Firefly TT.1 ...

Douglas AD-4W Skyraider

… und Douglas AD-4W Skyraider für Zielschleppaufgaben geleast. Erst mit der Indienststellung der Pilatus PC-6 Turbo Porter übernahmen eigene Flugzeuge des Österreichischen Bundesheeres diese Aufgabe.

… and Douglas AD-4W Skyraider for towing tasks. Only with the commissioning of the Pilatus PC-6 Turbo Porter own airplanes of the Austrian Armed Forces took over this task.

Sud Est S.E.3130 Alouette II

Im März 1958 kamen die ersten von insgesamt 16 Sud Est S.E.3130 Alouette II zum Österreichischen Bundesheer.

In March 1958 the first of a total of 16 Sud Est S.E.3130 Alouette II began their service with the Austrian Armed Forces.

Aufgrund ihrer Gipfelhöhe von 4000 Metern wurde die Sud Est S.E.3130 Alouette II als für den Einsatz im Hochgebirge geeignet angesehen.

Because their service ceiling of 4,000 metres (13,100 ft) the Sud Est S.E.3130 Alouette II was considered to be suitable for service in the high mountains.

Die Sud Est S.E.3130 Alouette II war unbewaffnet, konnte aber an einem Lasthaken unter dem Rumpf Lasten bis zu 600 Kilo transportieren.

The Austrian Sud Est S.E.3130 Alouette II was unarmed but could transport payloads up to 600 kilo (1,320 lb) on a hook under its fuselage.

Cessna L-19

Im Mai 1958 erhielt das Österreichische Bundesheer seine ersten Cessna L-19A. Die Flugzeuge wurden im Rahmen des Military Assistent Program (MAP) der Vereinigten Staaten zur Verfügung gestellt. Dieses beim Hersteller als Cessna Modell 305A bezeichnete, viersitzige Flugzeug wurde bei den amerikanischen Streitkräften als Beobachtungsflugzeug Cessna O-1 „Bird Dog", unter anderem auch in Vietnam, eingesetzt.

In May 1958, the Austrian Armed Forces got their first Cessna L-19A. These aircraft were on disposal as a part of the Military Assistant Program (MAP) from the United States. This type, named Cessna Model 305A by the manufacturer, was a four-seater and was used as an observation aircraft Cessna O-1 "Bird Dog" also in Vietnam.

In Österreich diente die Cessna L-19A als Verbindungsflugzeug, zur Schulung, etwa für Luftbildpersonal, aber auch …

In Austria the Cessna L-19A was used as liaison- and training airplane – among other usages also for aerial picture staff …

… als Zieldarsteller, wie hier für eine 40 mm Flugabwehrkanone Bofors 57 L 70 einer der beiden Flugabwehrbatterien des Österreichischen Bundesheeres.

... and as a training target, as here for a 40 mm anti-aircraft gun Bofors 57 L 70 one of the two Anti-aircraft batteries of the Austrian Armed Forces.

Das Österreichische Bundesheer betrieb bis 1997 insgesamt 22 Cessna L-19A und sieben L-19E. Die E-Version hatte den gleichen Motor, unterschied sich jedoch durch einen verstärkten Rumpf, stärkere Tragflächen, unter die zwei oder vier Pylons Mk.4A für Zielmarkierungsraketen montiert werden konnten, einen Verstellpropeller sowie einige weitere Details von der ersten Version der L-19A.

The Austrian Armed Forces operated until 1997 a total of 22 Cessna L-19A and seven L-19E. The E version had the same engine, but differentiated by a reinforced hull, stronger wings, under which two or four pylons Mk.4A could be mounted for target rockets, a variable pitch propeller and some more details from the first version of the L-19A.

Westland S-55 Whirlwind

Im Mai 1958 bestellte das Österreichische Bundesheer bei der Firma Westland Aircraft Ltd. in Yeovil, England, zehn Whirlwind Mk.2 für die 1. Hubschrauberstaffel in Linz-Hörsching.

In May 1958 the Austrian Armed Forces commissioned ten Whirlwind Mk.2, built by the Westland Aircraft Ltd. in Yeovil, Britain, for the 1st Helicopter Squadron in Linz-Hörsching.

Der Westland Whirlwind ist ein Lizenzbau der amerikanischen Sikorsky S-55 Chickasaw und kann Lasten bis 1040 Kilo, acht Soldaten oder auch Verwundete transportieren.

The Westland Whirlwind is a licensed construction of the American Sikorsky S-55 Chickasaw and can carry loads up to 1,040 kilos (2,290 lb), eight soldiers or wounded.

Während die anderen neun Whirlwind des Österreichischen Bundesheeres Militärversionen waren, handelte es sich bei der am 21. Juni 1956 als „4D-XQ“ registrierten Maschine um eine Zivilversion.

In contrast to the other nine Whirlwind of the Austrian Armed Forces, the helicopter with the registration “4D-XQ” (registered on 21 June 1956) was a civilian version.

Eine Szene vom Westland-Werksflugplatz in Yeovil, 1958: Während im Vordergrund eine kubanische Whirlwind startet, erkennt man im Hintergrund eine Maschine des Österreichischen Bundesheeres.

A scene from the Westland factory airfield in Yeovil, 1958: While in the foreground a Cuban Whirlwind takes off, one recognizes in the background a machine of the Austrian Armed Forces.

Piaggio P.149D

Ein Exemplar des Schulflugzeuges Piaggio P.149D mit der Werksnummer 344 wurde im Jahr 1958 vom Hersteller Industrie Aeronautiche e Meccaniche Rinaldo Piaggio SpA in Genua, Italien, dem Österreichischen Bundesheer geschenkt, wobei hinter dieser Idee wohl die Hoffnung auf einen Anschlussauftrag stand.

An example of the training aircraft Piaggio P.149D with the serial number 344 was donated in 1958 by the manufacturer Industrial Aeronautiche e Meccaniche Rinaldo Piaggio SpA in Genoa, Italy, the Austrian Armed Forces. Behind this idea probably stood the hope of a follow-up order.

Die Piaggio P.149D wurde neben der Pilotenschulung auch als Zielschleppflugzeug verwendet und Ende 1965 an einen privaten Halter verkauft.

Beside its task as a training aircraft the Piaggio P.149D was also used for target towing and at the end of 1965 it was sold to a private owner.

Fouga C.M.170R Magister

Da sich bald herausstellte, dass die De Havilland D.H.115 Vampire für die Anfangsschulung der Düsenpiloten des Österreichischen Bundesheeres nicht besonders geeignet war, wurde mit der Fouga C.M.170R Magister ab dem 12. Februar 1959 ein tauglicherer und kostengünstigerer Flugzeugtyp beschafft.

Since it soon turned out that the De Havilland D.H.115 Vampire was not particularly suitable for the initial training of jet pilots of the Austrian Armed Forces the Fouga C.M.170R Magister was procured from 12 February 1959, which was a more suitable and cheaper aircraft type.

Im Jahr 1962 stieg die Jabo-Schulstaffel von ihren acht De Havilland D.H.115 Vampire T.11 und T.55 auf 18 Fouga C.M.170R Magister um. Sechs dieser Düsentrainer hatten davor bei der Luftwaffe der Bundesrepublik Deutschland im Dienst gestanden.

In 1962 the Fighter bomber Training Squadron change their eight De Havilland D.H.115 Vampire T.11 and T.55 for 18 Fouga C.M.170R Magister. Six of these jet trainers had been in service of the German Air Force before.

Die Düsengrundschulung auf der Fouga Magister dauerte ungefähr 100 Stunden. Ab 1971 wurden die Düsentrainer ausgeschieden. Am 24. Juli 1972 erfolgte der letzte Flug einer österreichischen Magister.

The basic training on the Fouga Magister lasted about 100 hours. From 1971, these jet plane trainers were eliminated. On 24 July 1972, the last flight of an Austrian Magister took place.

Nach der Rückgabe der meisten Fouga Magister des Österreichischen Bundesheeres an den Hersteller Aérospatiale wurden vier dieser Maschinen von dort aus an das Irish Army Air Corps geliefert, und fünf C.M.170R gelangten von Österreich per Luftfracht direkt an das im Bürgerkrieg versunkene Biafra, was damals in den österreichischen Medien für Aufregung sorgte.

After the return of most of the Fouga Magister of the Austrian Armed Forces to the manufacturer Aérospatiale four of these machines were delivered from there to the Irish Army Air Corps, and five C.M.170R were sent by air directly to the war-torn Biafra, which then caused a stir in the Austrian media.

North American LT-6G Texan

Im Oktober 1959 trafen die ersten von zehn North American LT-6G Texan-Schulflugzeuge in Österreich ein und wurden im folgenden Jahr der Fliegerschulkompanie 2 in Linz-Hörsching zugeteilt.

In October 1959 the first of ten North American LT-6G Texan training airplanes arrived in Austria and in the following year these were allocated to the Fliegerschulkompanie 2 (Pilots Training Squadron 2) in Linz-Hörsching.

Als Pufferstaat zwischen den östlichen und westlichen Machtblöcken kam das neutrale Österreich nach der Wiederherstellung seiner staatlichen Souveränität wiederholt in den Genuss von Geschenken der beiden rivalisierenden Machtblöcke. Die North American LT-6G Texan waren ein Präsent der USA.

As a buffer state between the eastern and western power blocs, after the restoration of its sovereignty, neutral Austria repeatedly enjoyed the gifts of the two rival power blocs. The North American LT-6G Texan was a present from the USA.

Die Version LT-6G war ursprünglich für die bewaffnete Aufklärung im Koreakrieg entwickelt worden und konnte unter ihren Tragflächen Maschinengewehrbehälter, verschiedene ungelenkte Luft-Boden- sowie Rauchraketen und für Übungszwecke auch Zementbomben mitführen.

The LT-6G version was originally developed for armed reconnaissance in the Korean War and could carry under its wings machine gun containers, various unguided air-ground rockets and smoke flares and for practice purposes also cement bombs.

De Havilland Canada D.H.C. L-20 Beaver

Ein weiteres Geschenk der USA waren sechs De Havilland Canada D.H.C. L-20 Beaver, von denen die ersten fünf Maschinen am 28. Juli 1960, in Kisten verpackt, beim Aufklärungsausbildungskommando in Langenlebarn eintrafen. Die Lieferung der letzten Beaver erfolgte elf Monate später.

Another gift from the USA were six De Havilland Canada D.H.C. L-20 Beaver, of which the first five airplanes arrived in Langenlebarn on 28 July 1960, packed in boxes, at the reconnaissance training command. The delivery of the last Beaver took place eleven months later.

Die vom Hersteller als De Havilland Canada D.H.C.2 Beaver bezeichneten Flugzeuge, ehemalige Maschinen der United States Army Air Forces, dienten zur Blindflugschulung, für die Luftbildfotografie, zum Absetzen von Fallschirmspringern und wurden später, da die L-20 mit verstärktem Kabinenboden bis zu 680 Kilo Fracht befördern konnte, auch als Transportflugzeuge eingesetzt.

The airplane designated by the manufacturer as De Havilland Canada D.H.C.2 Beaver, former machines of the United States Army Air Force, were used for blind flight training, for aerial photography, for dropping parachutists and were later also used as a transport aircraft since the L-20 with a reinforced cabin floor could also carry up to 680 kilos (1,500 lb).

Nachdem die Hälfte der D.H.C. L-20 Beaver während ihres Einsatzes zerstört worden war, musterte das Österreichische Bundesheer die restlichen drei flugfähigen Maschinen Ende 1976 aus. Diese gelangten später in den Besitz von zivilen Betreibern in Alaska.

After half of the D.H.C. L-20 Beaver had been destroyed during their mission, the Austrian Armed Forces surveyed the remaining three airworthy aircraft end of 1976. These later came into the hands of civilian operators in Alaska.

Bell H-13H Sioux

Noch bevor die ersten von 17 Bell H-13H, ein Geschenk der USA, Mitte Oktober 1960 in Österreich eintrafen, wurde in Aigen im Ennstal bereits die zweite Hubschrauberstaffel aufgestellt.

Even before the first of 17 Bell H-13H, a gift from the USA, arrived in Austria in mid-October 1960, the second helicopter squadron was set up in Aigen in the Enns Valley.

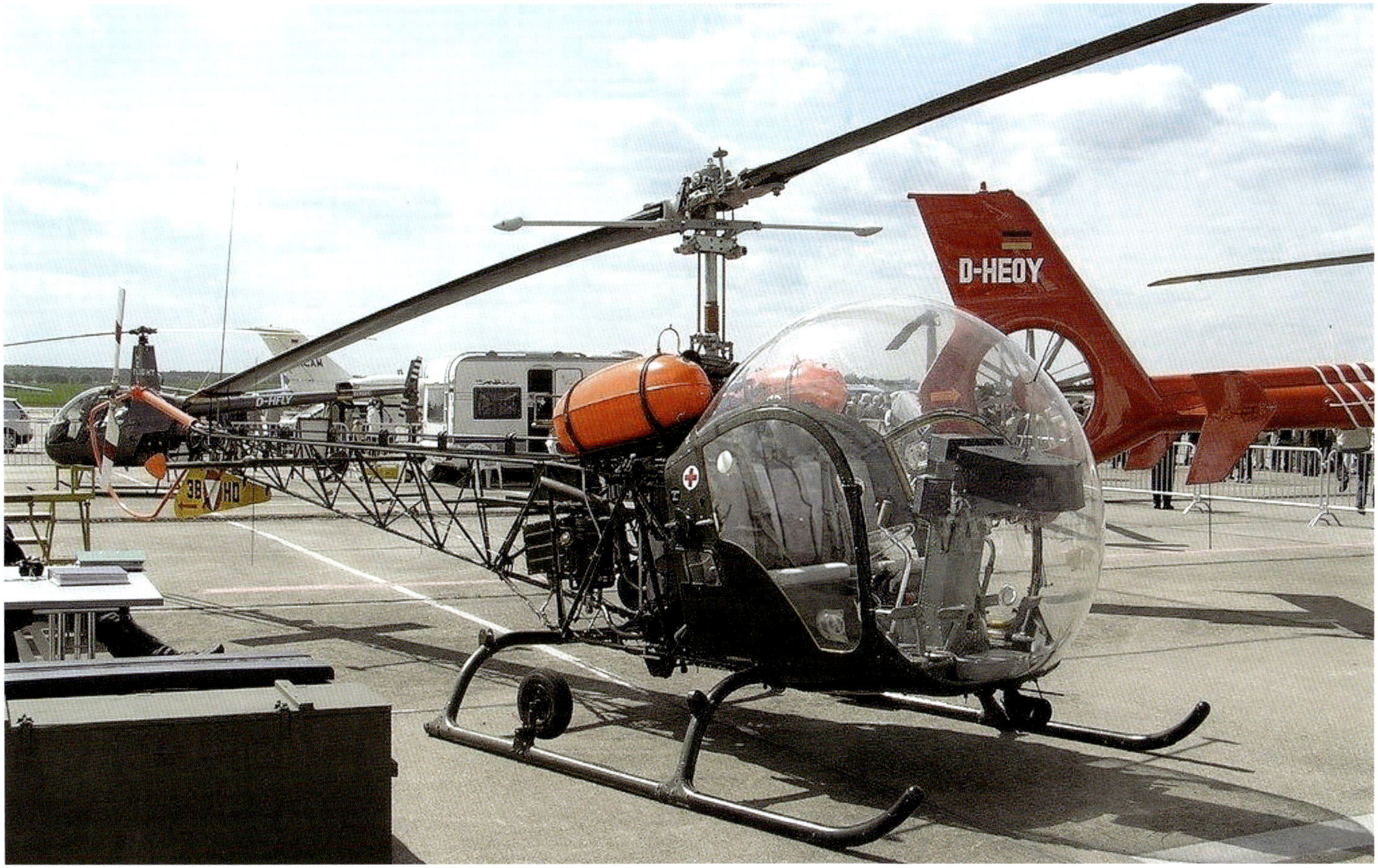

Bell H-13H Sioux lautet die Bezeichnung dieses Hubschraubers bei den amerikanischen Streitkräften. Seine Leistungen waren mit jenen der Agusta-Bell 47G2 nahezu identisch.

Bell H-13H Sioux is the name of this helicopter in the US Armed Forces. His performance was almost identical to those of the Agusta Bell 47G2.

Im Jahr 1964 führten die Bell H-13H des Österreichischen Bundesheeres ihren 200. Rettungsflug durch, doch erwies sich dieser Typ für Einsätze im Hochgebirge als zu schwach, weshalb diese verstärkt mit der Sud Est SE-3130 Alouette II durchgeführt wurden.

In 1964, the Austrian Armed Forces' Bell H-13H made its 200th rescue flight, but this type proved to be too weak for use in the high mountains, which is why these operations were increasingly carried out with the Sud Est SE-3130 Alouette II.

Am 8. Oktober 1976 wurde der letzte Bell H-13H des Österreichischen Bundesheeres mit der Registrierung „3B-HN“ außer Dienst gestellt und an die Höhere Technische Lehranstalt in Hollabrunn abgegeben.

On 8 October1976, the last Bell H-13H of the Austrian Armed Forces with the registration “3B-HN” was decommissioned and delivered to the Higher Technical College in Hollabrunn.

Saab J29F Tonne

Nachdem die österreichische Regierung 1960 die Beschaffung von Abfangjägern beschlossen hatte, trafen im Juli 1961 die ersten von insgesamt 30 Saab J29F in Wien-Schwechat ein.

After the Austrian government had decided in 1960 to purchase interceptors, the first of a total of 30 Saab J29F arrived in July 1961 in Vienna-Schwechat.

Die Saab J29F war als Jagdbomber ausgelegt. Wegen ihrer Form bekam die J29 bald den Spitznamen „Tunnan“ oder „Fliegende Tonne“.

The Saab J29F was a fighter-bomber and because of its shape received the nickname “Tunnan” or “Flying Barrel”.

Während des „Prager Frühlings“ und des deshalb folgenden Einmarschs von Truppen des Warschauer Pakts in die Tschechoslowakei in der Nacht vom 21. auf den 22. August 1968 waren zur Grenzsicherung ständig zwei Flugzeuge und/oder Hubschrauber des Österreichischen Bundesheeres entlang der Grenze zur ČSSR auf Patrouille. Den Großteil dieser Grenzsicherungsflüge führten jedoch die Saab J29F durch, die innerhalb von 24 Stunden insgesamt 96 Einsätze absolvierten.

During the "Prague Spring" and the subsequent invasion by the Warsaw Pact troops into Czechoslovakia on the night of 21 to 22 August 1968, two airplanes and / or helicopters of the Austrian Armed Forces along the Czechoslovak border were constantly on border patrol. However, most of these border security flights were carried out by the Saab J29F, which completed a total of 96 missions within 24 hours.

Die Saab J29F war mit vier 20 mm Akan M/47C bewaffnet und konnte als Jagdbomber unter den Tragflächen acht ungelenkte 6,3 oder 7,5 cm Raketen oder an deren Stelle zwei 450 Liter fassende Kerosintanks mitführen. Von der Svenska Flygverkstäderna in Malmö wurden zusätzlich zehn Fotoaufklärer-Umrüstsätze beschafft, die anstelle der beiden linken Bordkanonen montiert werden konnten und über je drei 70-mm-Vinten-Kameras verfügten.

The Saab J29F was armed with four 20 mm Akan M/47C and could fly eight unguided 6.3 or 7.5 cm rockets as fighter-bombers under the wings or two 450 litre kerosene tanks in their place. In addition, Svenska Flygverkstäderna in Malmö procured ten photo reconnaissance conversion kits, which could be mounted in place of the two left-hand cannons and each had three 70-mm Vinten cameras.

Bis zu ihrer Außerdienststellung im Juli 1972 führten die Saab J29F des Österreichischen Bundesheeres 18.310 Einsätze durch und waren dabei insgesamt 13.205 Stunden und acht Minuten in der Luft.

Until their retirement in July 1972, the Saab J29F of the Austrian Armed Forces carried out 18,310 missions, leaving a total of 13,205 hours and eight minutes in the air.

Saab 91D Safir

Um die Pilotenausbildung beim Österreichischen Bundesheer zu vereinfachen und die verschiedenen Schulflugzeugtypen – auch „Schmetterlingssammlung" genannt – zu verringern, wurden ab dem 1. August 1964 insgesamt 24 Saab 91D Safir in Dienst gestellt.

In order to simplify the pilot training at the Austrian Armed Forces and to reduce the various types of trainer aircraft – also called "butterfly collection" – a total of 24 Saab 91D Safir were put into service from 1 August 1964.

Die Saab 91D Safir, die letzte Version dieses in mehr als 20 Länder gelieferten schwedischen Verkaufsschlagers, sollte ursprünglich Saab 91 Staffett heißen. Sie wurde von dem schwedischen Konstrukteur Anders J. Anderson entwickelt, der davor in Deutschland bei den Bücker Werken gearbeitet hatte.

The Saab 91D Safir, the latest version of this Swedish best seller delivered in more than 20 countries was originally named Saab 91 Staffett. It was developed by Swedish designer Anders J. Anderson, who had previously worked in Germany at the Bücker plants.

Als die Grundschulung der Bundesheerpiloten ab 1989 von der neuen Pilatus PC-7 Turbo Trainer übernommen wurde, verwendete man die verbleibenden Saab Safir in abnehmender Anzahl bis zur Außerdienststellung der letzten Saab 91D am 13. September 1993 vermehrt für Verbindungsaufgaben.

When the basic training of the Armed Forces pilots was taken over in 1989 by the new Pilatus PC-7 Turbo Trainer, the remaining Saab Safir were used in decreasing numbers until the decommissioning of the last Saab 91D on 13 September 1993 but increasingly for liaison tasks.

Nachdem noch im Jahr ihrer Außerdienststellung acht ausgeschiedene Saab Safir in private Hände kamen, gründeten sieben Bundesheerpiloten mit diesem Schulflugzeugtyp die Kunstflug-Viererformation „Team 2000“, um auf Flugveranstaltungen das Publikum zu begeistern.

After eight retired Saab Safir came into private hands in the year of their decommissioning, seven federal army pilots founded the aerobatic four-formation "Team 2000" with this type of training aircraft to inspire the public at air events.

Sud Aviation S.A.316B Alouette III

Am 28. Jänner 1967 trafen in Österreich die ersten vier von insgesamt 28 angekauften Mehrzweckhubschraubern Sud Aviation S.A.316B Alouette III ein, die für Transport-, Sanitäts- und Feuerlöschaufgaben eingesetzt werden können und bis zum heutigen Tag im Dienst stehen.

On 28 January 1967, the first four of a total of 28 purchased multi-purpose helicopters Sud Aviation S.A.316B Alouette III arrived in Austria and can be used for transport, medical and fire-fighting tasks and are still in service to this day.

In den letzten Jahrzehnten bewährten sich die Alouette III sowohl bei Ausbildungs- und Transportaufgaben für das Österreichische Bundesheer als auch bei Hochwasser- und Lawineneinsätzen sowie Rettungsflügen im Hochgebirge und der Waldbrandbekämpfung.

In recent decades, the Alouette III have proven themselves in training and transport tasks for the Austrian Armed Forces as well as in flood and avalanche rescue operations as well as rescue flights in the high mountains and forest fire fighting.

Bis zur Übernahme von notärztlichen Flügen durch den ÖAMTC (Österreichischer Automobil-, Motorrad- und Touring Club) ab dem 1. Juli 2001 waren zwei speziell für solche Einsätze ausgerüstete Alouette III in das österreichische Notarzthubschrauber-System eingebunden und standen täglich von Sonnenaufgang bis -untergang für Notfälle bereit.

Until the takeover of emergency flights by the ÖAMTC (Austrian Automobile, Motorcycle and Touring Club) from 1 July 2001, two Alouette III specially equipped for such operations were involved in the Austrian emergency medical helicopter system, available every day from sunrise to sunset ready for emergencies.

Als Österreich am 1. Dezember 1997 die Verantwortung für die Schengen-Außengrenze übernahm, wurde dem Hubschraubergeschwader in Aigen im Ennstal ein 1250 Kilometer langer Grenzabschnitt, von Kittsee gegenüber Bratislava bis zur Soboth in der Steiermark, übertragen. Dafür wurden die Sud Aviation S.A.316B mit moderner Überwachungs- und Kommunikationstechnik ausgerüstet. Sie flogen jährlich ungefähr 1000 Stunden für die Grenzüberwachung.

When Austria assumed responsibility for the Schengen external border on 1 December 1997, the helicopter squadron in Aigen in the Enns Valley was given a 1,250 kilometre border section from Kittsee across Bratislava to Soboth in Styria. For this, the Sud Aviation S.A.316B was equipped with modern monitoring and communication technology. They flew about 1,000 hours per year for border surveillance.

Agusta-Bell AB 204B

Die Agusta-Bell AB 204B ist ein italienischer Lizenzbau der Bell 204, welche bei den amerikanischen Streitkräften als UH-1 Iroquois bezeichnet wird. Die AB 204B des Österreichischen Bundesheeres ist mit einer Seilwinde und einem Außenlasthaken ausgerüstet und kann drei Krankentragen samt medizinischem Personal transportieren.

The Agusta-Bell AB 204B is an Italian licensed construction of the Bell 204, which is referred to in the US armed forces as UH-1 Iroquois. The AB 204B of the Austrian Armed Forces is equipped with a winch and an outside hook and can carry three stretchers and medical personnel.

Im Juni 1963 trafen die ersten von insgesamt 26 Agusta-Bell AB 204B in Österreich ein. Sie waren für die beiden Staffeln des Hubschraubergeschwaders in Langenlebarn und Linz-Hörsching vorgesehen.

In June 1963, the first of a total of 26 Agusta-Bell AB 204B arrived in Austria. They were intended for the two squadrons of the helicopter squadron in Langenlebarn and Linz-Hörsching.

Die Ausmusterung der AB 204B war bis zum Oktober 1982 vorgesehen, doch nachdem im Mai 1981 die beiden Großraumhubschrauber Sikorsky S-65OE überraschend an Israel verkauft worden waren, fehlte dem Hubschraubergeschwader nun Transportkapazität. Da nur neun Agusta-Bell AB 204B verkauft werden konnten, wurde, basierend auf der Ersatzteillage, der erste der verbliebenen Hubschrauber im selben Jahr wieder in Dienst gestellt. Bis zum Mai 1985 folgten diesem weitere sieben der 1982 ausgemusterten AB 204B.

The AB 204B was scheduled to be decommissioned by October 1982, but after the two large-capacity helicopters Sikorsky S-65OE had been surprisingly sold to Israel in May 1981, the helicopter squadron now lacked transport capacity. Since only nine Agusta-Bell AB 204B could be sold, the first of the remaining helicopters was reinstated in the same year, based on the spare parts situation. By May 1985, further seven of the 1982 decommissioned AB 204B followed.

Mitte der 1980er-Jahre übernahm der größere Agusta-Bell AB 212 die Aufgaben des AB 204B. Der letzte dieser acht Hubschrauber wurde Ende Juni 2001 endgültig außer Dienst gestellt.

In the mid-1980s, the larger Agusta-Bell AB 212 took over the tasks of the AB 204B. The last of these eight helicopters were finally decommissioned at the end of June 2001.

Agusta-Bell AB 206A Jet Ranger

Da sich Ende der 1960er-Jahre die Dienstzeit der Agusta-Bell 47G2 ihrem Ende näherte und auch jenes der Bell H-13H und der Sud Est S.E.3130 Alouette II absehbar war, wurde nach einem Nachfolger dieser Hubschraubertypen gesucht. Gegen die Konkurrenzmodelle Fairchild-Hiller FH-1100 (OH-5A) und Hughes 500 (OH-6A Cayuse) setzte sich die Bell 206A Jet Ranger (OH-4A) durch.

Since by the end of the 1960s, the service time of the Agusta Bell 47G2 was nearing its end and also that of the Bell H-13H and the South Est S.E.3130 Alouette II were foreseeable, a successor to these types of helicopters had to be found. The Bell 206A Jet Ranger (OH-4A) prevailed over its competitors Fairchild-Hiller FH-1100 (OH-5A) and Hughes 500 (OH-6A Cayuse).

Zwischen Juni 1969 und Oktober 1970 wurden 13 bei der Costruzioni Aeronautiche Giovanni Agusta SpA in Cascina Costa, Italien, gebaute AB 206A Jet Ranger nach Langenlebarn geliefert, wo sie primär zur Grundschulung verwendet wurden, aber auch mit zwei Krankentragen Verwundeten- und Lasttransporte bis circa 500 Kilo durchführten.

Between June 1969 and October 1970, 13 AB 206A Jet Rangers were built by the Costruzioni Aeronauticals Giovanni Agusta SpA in Cascina Costa, Italy and were delivered to Langenlebarn, where they were primarily used for basic training but equipped with two stretchers to carry wounded and transport loads of up to 500 kilos (1,102.31 lb).

Nach ungefähr 100.000 Flugstunden in 40 Jahren erfolgte bei der „Airpower 2009“ in Zeltweg das „Fly-out“ der Agusta-Bell AB 206A Jet Ranger des Österreichischen Bundesheeres. Zehn der verkauften AB 206A gelangten schließlich nach Australien.

After about 100,000 flight hours in 40 years, the Agusta-Bell AB 206A Jet Ranger of the Austrian Armed Forces were honoured with the "fly-out" at the "Air-Power 2009" in Zeltweg. Ten of the sold AB 206A finally arrived in Australia.

Die 2. Staffel des Hubschraubergeschwaders 1 in Langenlebarn wurde auch für zivile Assistenzeinsätze herangezogen und stellte Piloten und Equipment für das Hubschrauber-Kunstflugteam „Kleeblatt".

The 2nd Squadron of the Helicopter Squadron 1 in Langenlebarn was also used for civilian Assistance Missions and provided pilots and equipment for the helicopter aerobatic team Shamrock.

Short SC.7 Skyvan

In den 1960er-Jahren bestand kurzzeitig die Absicht, zum Zweck der Typenbereinigung der österreichischen „Schmetterlingssammlung“ 15 bis 20 zu dieser Zeit in den USA relativ preiswert angebotene De Havilland Canada D.H.C.3 Otter zu beschaffen. Als im Jahr 1966 zur Entscheidungsfindung des künftigen Flugzeugtyps Präsentationen von De Havilland D.H.C.2 Turbo Beaver, D.H.C.6 Twin Otter, Pilatus PC-6 Turbo Porter und Short SC.7 Skyvan durchgeführt wurden, fiel die Entscheidung zugunsten der Skyvan aus.

In the 1960s there existed briefly the intention to purchase for the purpose of streamlining the Austrian "Butterfly Collection" 15 to 20 De Havilland Canada D.H.C.3 Otter, which were offered at this time relatively cheaply in the United States. In 1966, when presentations of De Havilland D.H.C.2 Turbo Beaver, D.H.C.6 Twin Otter, Pilatus PC-6 Turbo Porter and Short SC.7 Skyvan were made in order to decide on the future type of airplane, the decision was made in favour of Skyvan.

Aus Kostengründen wurden am 23. Oktober 1968 allerdings nur zwei Short SC.7-3M-400 Skyvan bestellt, die im September des folgenden Jahres in Langenlebarn eintrafen und der 3. Staffel zugewiesen wurden. Aufgrund der geringen Anzahl nutzte man die beiden Short Skyvan intensiv. Ihre hauptsächlichen Destinationen waren neben den Nachbarländern Österreichs auch die Nachfolgestaaten von Jugoslawien sowie Schweden.

For cost reasons, however, only two Short SC.7-3M-400 Skyvan were ordered on 23 October 1968, which arrived in September of the following year in Langenlebarn and were assigned to the 3rd Squadron. Due to its small number the two Shorts Skyvan were used too intensively. Its main destinations, apart from Austria‘s neighbouring countries, were the successor states of Yugoslavia and Sweden.

Anfang April 2007 erging der Befehl, die beiden Short SC.7 „auszuphasen“. Nach 32.059 Flügen, bei denen die beiden STOL-Transportflugzeuge (Short Take-Off and Landing) 16.182 Flugstunden durchgeführt und neben dem Transport von 36.000 Passagieren auch 100.000 Fallschirmspringer abgesetzt hatten, flog die Skyvan „5S-TA“ (WNr. 1855) am 30. November 2007 das letzte Mal, bevor sie der Militärluftfahrtsammlung in Zeltweg übergeben wurde. Die andere Skyvan „5S-TB“ (WNr. 1860) war einen Tag davor an die dänische DAO Aviation A/S verkauft worden, welche sie nach einigen Umbauten an die Swala Aviation in der Volksrepublik Kongo weiterverkaufte.

At the beginning of April 2007, the order was issued to “phase out” the two Short SC.7. After 32,059 flights in which the two STOL (short take-off and landing) transport aircraft carried out 16,182 flight hours and dropped 100,000 parachutists in addition to transporting 36,000 passengers, the Skyvan “5S-TA” (WNr. 1855) flew for the last time on 30 November 2007 before being handed over to the Military Aviation Collection in Zeltweg. The other Skyvan “5S-TB” (No. 1860) had been sold the day before to the Danish DAO Aviation A/S, which resold it after some modifications to the Swala Aviation in the People’s Republic of Congo.

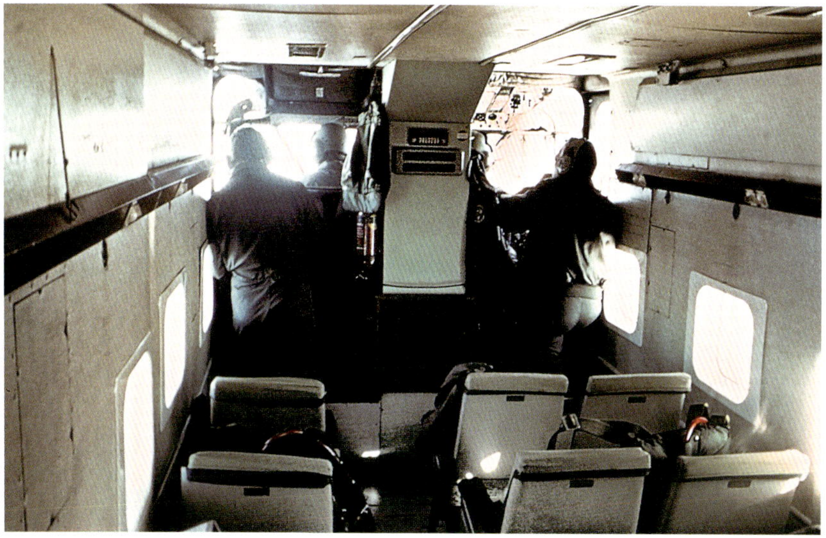

Saab 105OE

Seite/page 88

Ende der 1960er-Jahre wurde es notwendig, einen Nachfolger für das in die Jahre gekommene Jagdflugzeug Saab J29F Tonne und den Düsentrainer Fouga C.M.170R Magister zu finden. Ursprünglich war geplant, 20 Saab 105XT, welche für Grundschul-, Erdkampf- und Aufklärungsaufgaben entwickelt worden waren, als Nachfolger der „Tonne" zu erwerben und weitere Maschinen der Exportversion Saab 105XT bei Bedarf nach und nach als Ersatz für die in den kommenden Jahren auszuscheidenden Fouga Magister zu beschaffen. Das abgebildete Foto entstand auf dem „Aerosalón" 1971 in Le Bourget und zeigt einen Demonstrator Saab XT der Svenska Aeroplan Aktiebolaget AB in den nicht ganz korrekten Markierungen der Österreichischen Luftstreitkräfte.

In the late 1960s, it became necessary to find a successor to the outdated fighter Saab J29F Tonne and the jet trainer Fouga C.M.170R Magister. Originally it was planned to acquire 20 Saab 105XT, which had been developed for basic training, ground combat and reconnaissance tasks, as the successor of the "Tunnan" and other machines of the export version Saab 105XT as needed to replace over the next few years the Fouga Magister, which was due to be decommissioned. The picture shown was taken at the "Aerosalón" 1971 in Le Bourget and shows the Saab XT (demonstration) of the Svenska Aeroplan Aktiebolaget AB in the not quite correct markings of the Austrian Air Force.

Seite/page 89

Am 2. Juli 1970 trafen die ersten Saab 105OE in Linz-Hörsching ein. Die Gesamtzahl der erworbenen Maschinen stieg in den folgenden Jahren auf 40 an. Da es Österreich aufgrund des permanenten Unverständnisses der politischen Opposition in diesen mehr als eineinhalb Jahrzehnten an echten Abfangjägern mangelte, mussten die Düsentrainer auch zur Luftraumüberwachung herangezogen werden. Das bedeutete, dass bis zur Beschaffung der Saab Draken die schnellsten österreichischen Flugzeuge die Douglas DC-9 der Austrian Airlines waren.

On 2 July 1970, the first Saab 105OE arrived in Linz-Hörsching. The total number of purchased machines rose to 40 in the following years. Since Austria lacked real interceptors due to the permanent lack of understanding of the political opposition in these more than one and a half decades, the jet trainers also had to be involved in airspace surveillance. This meant that until the acquisition of Saab Draken the fastest Austrian aircraft was the Douglas DC-9 of Austrian Airlines.

GEFAHR
GEFAHR

Seite/page 91

Am 12. Juli 1975 wurde erstmals der aus Piloten und Maschinen der 2. Staffel des Jabo-Geschwaders in Graz-Thalerhof gebildete Kunstflugschwarm „Karo As“ der Öffentlichkeit vorgestellt. Zwei Jahre später errangen die vier Saab 105OE mit ihrer Vorführung bei der „Shell Trophy“ in England den zweiten Platz im militärischen Kunstflug und nahmen in den folgenden Jahren erfolgreich an weiteren Einzel- und Formationskunstflugbewerben teil. Das Foto zeigt eine Saab 105OE im Anstrich des ehemaligen Kunstflugteams „Karo As“, aufgenommen 1990 in Innsbruck-Kranebitten.

On 12 July 1975 pilots and machines of the 2nd Squadron of the Fighter-Bomber squadron in Graz-Thalerhof, who formed the aerobatic team “Karo As” (“Ace of Spades”), were presented for the first time to the public. Two years later, the four Saab 105OE won second place in military aerobatics with their performance at the “Shell Trophy” in England and successfully participated in further individual and formation aerial competitions in the following years. The photo shows a Saab 105OE in the painting of the former aerobatic team “Karo As”, taken in 1990 in Innsbruck-Kranebitten.

Seite/page 93

Nach der Einführung der Saab 105OE beim Österreichischen Bundesheer stieg 1970 auch der Kunstflugschwarm „Silver Birds“ von den ausgeschiedenen Fouga C.M.170R Magister auf den schwedischen Düsentrainer um und landete in den folgenden Jahren bei internationalen Kunstflugwettbewerben wiederholt auf den vorderen Plätzen. Heute stehen nur mehr zwölf Saab 105OE im Einsatz. Neben einer Reihe von Verlusten durch Flugunfälle in den letzten Jahrzehnten wurden sechs Maschinen aus dem Einsatz genommen, die übrigen dienen als Ersatzteilspender für das letzte Dutzend noch einsatzfähiger Saab 105OE in Österreich.

After the introduction of the Saab 105OE to the Austrian Armed Forces the aerobatic team “Silver Birds” changed in 1970 from the decommissioned Fouga C.M.170R Magister to the Swedish jet trainer and landed in the following years at international aerobatics competitions repeatedly in front positions. Today, only twelve Saab 105OE are in use. In addition to a series of aircraft accident losses in recent decades, six aircraft have been withdrawn from service, with the remainder serving as spare parts dispensers for the last dozen still operational Saab 105OE in Austria.

B
F
J
H
E

Sikorsky S-65OE

Seite/page 94

Ab 1958 betrieb das Österreichische Bundesheer eine gemischte Flotte aus kleineren und mittleren Hubschraubertypen. Nach den Hochwasserkatastrophen in den Jahren 1965 und 1966 wurde der Entschluss gefasst, für künftige Naturkatastrophen in jedem der neun Bundesländer Österreichs einen Großhubschrauber zu stationieren. Nachdem die Konkurrenzmuster Agusta A-101G, Boeing CH-47 Chinook, Mil Mi-8 „Hip“ bereits im Vorfeld ausgeschieden waren, fiel die Entscheidung gegen den einzigen ernsthaften Mitbewerber, den Sud Aviation S.A.321 Super Frelon, zugunsten des leistungsstärkeren Sikorsky S-65C-2, der Exportversion des CH-53C Sea Stallion, den die US-Marines als Erstkunde erhalten hatten und der während des Vietnamkrieges in allen Varianten den Spitznamen „Super Jolly“ bekam.

From 1958 the Austrian Armed Forces operated a mixed fleet of small and medium helicopter types. After the flood disasters in 1965 and 1966, the decision was made to station a large helicopter for future natural disasters in each of the nine federal states of Austria. After the competitors Agusta A-101G, Boeing CH-47 Chinook and Mil Mi-8 "Hip" had already been eliminated in advance, the decision was made against the only serious competitor, the Sud Aviation SA321 Super Frelon, in favour of the more powerful Sikorsky S- 65C-2, the export version of the CH-53C Sea Stallion, which had first been delivered to the US-Marines and had earned the nickname "Super Jolly" during the Vietnam War in all its variations.

Seite/page 96

Diese in Österreich als Sikorsky S-65OE bezeichneten Großhubschrauber erhielten laut dem Heeresluftfahrzeugregister ihre Markierungen am 9. November 1970 und wurden der 2. Staffel des Hubschraubergeschwaders 3 in Linz-Hörsching zugewiesen. Der S-65OE konnte neben seiner dreiköpfigen Besatzung wahlweise 38 Passagiere oder 22 Krankentragen plus vier Personen medizinischen Personals, intern mehr als 6000 Kilo Fracht oder – bei verringerter Treibstoffmenge – am Außenlasthaken bis zu 8000 Kilo Zuladung mitführen. Da beide Hubschrauber eine Höhe von knapp 6000 Meter Flughöhe erreichten, war ihr Einsatz im Gebirge kein Problem.

These large helicopters, designated as Sikorsky S-65OE in Austria, received their markings on 9 November 1970, according to the Army Aviation Register, and were assigned to the 2nd Squadron of Helicopter Regiment 3 in Linz-Hörsching. In addition to its crew of three, the S-65OE was able to carry 38 passengers or 22 stretchers plus four medical personnel, more than 6,000 kilos of freight internally, or up to 8,000 kilos of cargo at the outer cargo hook, with reduced fuel. Since both helicopters reached a height of almost 6,000 metres altitude, their use in the mountains was no problem.

5L MB

Da für die Ausrüstung des Österreichischen Bundesheeres wieder einmal das Geld fehlte, war die Beschaffung von neun dieser Sikorsky S-65OE reines Wunschdenken. Es blieb bei den beiden Großhubschraubern. Erschwerend kam hinzu, dass die beiden Sikorsky S-65OE immer wieder von technischen Problemen geplagt wurden und der Ankauf von Ersatzteilen sehr langwierig war. Durch ihre langen Standzeiten und das Ausbleiben von Naturkatastrophen, für die man die S-65OE eigentlich beschafft hatte, wurden diese beiden Hubschrauber nicht oft benötigt. Sie absolvierten pro Jahr nur zwischen 82 und 107 Flugstunden.

Since there was once again no money to buy equipment for the Austrian Armed Forces the purchase of nine of these Sikorsky S-65OE was pure wishful thinking. It should stay with the two big helicopters. To make matters worse, the two Sikorsky S-65OE were always plagued by technical problems and the purchase of spare parts was very expensive. Because of their long downtimes and the absence of natural disasters for which the S-65OE had actually been procured, these two helicopters were not often needed. They were only in the air between 82 and 107 hours per year.

Das Bundesministerium für Landesverteidigung beschloss daher, die beiden Sikorsky S-65OE ohne vorherige Ausschreibung zu verkaufen. Für einen Gesamtpreis von 52,8 Millionen Schilling gingen die Hubschrauber inklusive der Ersatzteile und Spezialwerkzeuge am 3. Mai 1981 an Israel – was nach Aussage des Herstellerwerkes Sikorsky Aircraft, Division of United Technologies Corporation, nur etwa einem Drittel ihres Wertes entsprach. Einer der beiden Sikorsky S-65, nun in Diensten der Heyl Ha'Avir, wurde am 31. März 1996 bei einem Startunfall zerstört.

The Federal Ministry for Defence therefore decided to sell the two Sikorsky S-65OE without prior call for tender. For a total price of 52.8 million schillings (€ 3.83 million), the helicopters including the spare parts and special tools went to Israel on 3 May 1981 – which, according to the manufacturer Sikorsky Aircraft, Division of United Technologies Corporation, was only about a third of their value. One of the two Sikorsky S-65, now in the service of Heyl Ha'Avir, was destroyed on 31 March 1996 during a start.

Bell OH-58B Kiowa

Trotz der Beschaffung der 13 Agusta-Bell AB 206A Jet Ranger in den Jahren 1969 und 1970 zeigte sich beim Österreichischen Bundesheer Mitte der 1970er-Jahre ein großer Mangel an Schul-, Aufklärungs- und Verbindungshubschraubern. Da aber nicht alle der für die Beschaffung von Fluggerät zuständigen Stellen dies akzeptierten, kam es vorerst zu keinem Nachkauf von weiteren zwölf Jet Rangern. Im Jahr 1976 fiel dann eher überraschend die Entscheidung, anstelle der vorgeschlagenen AB 206A ein Dutzend Bell OH-58B Kiowa samt dem Waffensystem General Electric 7.62 mm M27 Minigun zu erwerben.

Despite the purchase of the 13 Agusta-Bell AB 206A Jet Ranger in the years 1969 and 1970, the Austrian Armed Forces suffered from a great shortage of schooling, reconnaissance and liaison helicopters in the mid-1970s. Since not all of those responsible persons for the procurement of aircraft accepted this, no additional purchases for the time being of the planned twelve Jet Rangers were made. Then, in 1976, it was rather a surprise decision, a dozen Bell OH-58B Kiowa and the General Electric 7.62 mm M27 Minigun weapon system were bought instead of the proposed AB 206A.

Die erste Hälfte der bestellten OH-58B wurde am 14. Juni 1976 per Bahntransport in Langenlebarn angeliefert, acht Tage später hob Hauptmann Michel erstmals mit einem österreichischen Kiowa ab. Nach dem Zusammenbau der zweiten Tranche, die am 16. September 1976 in Langenlebarn eingetroffen war, wurden alle Bell OH-58B der 3. Staffel des Hubschraubergeschwaders 1 zugeteilt.

The first half of the ordered OH-58B was delivered on 14 June 1976 by rail transport in Langenlebarn. Eight days later Captain Michel took off for the first time with an Austrian Kiowa. After the assembly of the second tranche, which had arrived in Langenlebarn on 16 September 1976, all Bell OH-58B were assigned to the 3rd Squadron of the Helicopter Air Regiment.

Teil des Waffensystems M27 der Bell OH-58B ist die sechsläufige 7,62 mm Minigun M134, die pro Minute 2000 oder 4000 Schuss auf Ziele in bis zu 1000 Meter Entfernung abgeben kann. Für Fotoflüge erhält der Kiowa ein spezielles Fotofenster montiert.

Part of the M27 weapon system of the Bell OH-58B is the six-barreled 7.62 mm Minigun M134, which can deliver 2,000 or 4,000 rounds per minute to targets up to 1,000 metres away. For photo flights, the Kiowa has a special photo window mounted.

Ab 1997 wurden die OH-58B auch für die Überwachung der neuen Schengen-Außengrenzen herangezogen. Um die Bodenfreiheit für neue Navigations- und Steuergeräte zu gewährleisten, erhielten alle Hubschrauber ab dem Juni desselben Jahres höhere Kufen. Der Bell OH-58B Kiowa ist der einzige bewaffnete Hubschrauber des Österreichischen Bundesheeres.

From 1997, the OH-58B was also used to monitor the new external Schengen borders. In order to ensure ground clearance for new navigation and control devices, all helicopters were upgraded with higher skids from June of the same year. The Bell OH-58B Kiowa is the only armed helicopter of the Austrian Armed Forces.

Pilatus PC-6 Turbo Porter

Obwohl im Jahr 1966 bei einer Evaluierung von Transportflugzeugen der Short SC.7 Skyvan gegenüber der Pilatus PC-6 Turbo Porter der Vorzug gegeben wurde, bot die Pilatus Flugzeugwerke AG in Stans acht Jahre später dem Österreichischen Bundesheer die Erprobung ihrer PC-6 als Zielsack-Schleppflugzeug an. Im folgenden Jahr wurde dieses STOL-Flugzeug daher in Oggau von der leichten Fliegerabwehr getestet. Als Ergebnis des Tests bestellte das Österreichische Bundesheer im Februar 1976 eine dieser Maschinen.

Although in 1966 in an evaluation of transporters the Short SC.7 Skyvan was given preference over the Pilatus PC-6 Turbo Porter, Pilatus Flugzeugwerke AG in Stans offered the Austrian Armed Forces eight years later the opportunity to test their PC-6 as a Target Bag-Towing aircraft. In the following year, this STOL aircraft (Short Take-Off and Landing) was therefore tested in Oggau by the light antiaircraft artillery. As a result of the test, the Austrian Armed Forces ordered one of these machines in February 1976.

Der Erstbestellung folgte im April 1976 ein weiterer Auftrag über elf zusätzliche Turbo Porter. Alle zwölf Pilatus PC-6 wurden der 4. Flächenstaffel des Fliegerregiments 1 in Langenlebarn zugeteilt. Bei Bedarf konnten die Turbo Porter mit Skiern ausgerüstet werden. Neben der Verwendung als Zielsack-Schleppflugzeuge mit einem 1000 bis 1500 Meter langen Seil kamen sie als Fracht- und Krankentransporter, Verbindungsflugzeuge, für Luftbildaufnahmen und Vermessungsflüge, zum Absetzen von Fallschirmjägern sowie als Löschflugzeuge mit einem 800 Liter fassenden, bei Bedarf in der Kabine montierbaren Wassertank zum Einsatz.

The initial order was followed by another one for eleven additional Turbo Porter in April 1976. All twelve Pilatus PC-6 were assigned to the 4th Squadron of the 1st Air Regiment in Langenlebarn. If needed, the Turbo Porter could be equipped with skis. In addition to being used as target bag towing aircraft with a 1,000 to 1,500 metres rope, they were used as cargo and ambulance transporters, liaison aircraft, for aerial photography and surveying flights, deployment of paratroopers and fire fighting aircraft with a 800 litre water tank mountable in the cabin if required.

Als Kurzstart- und Landeflugzeug kann die Pilatus PC-6 Turbo Porter in unwegsamem und mit ihrer Dienstgipfelhöhe von 10.000 Metern auch in hochalpinem Gelände eingesetzt werden. Am 18. August 2004 übernahm das Österreichische Bundesheer vom Innenministerium eine 13. Pilatus PC-6 Turbo Porter, die davor für das Bundesamt für Eich- und Vermessungswesen als Kalibrator geflogen worden war. Der blaue Anstrich dieses Flugzeugs wurde nach der Übernahme durch die 4. Staffel des Fliegerregiments 1 beibehalten.

As a short-range take-off and landing airplane, the Pilatus PC-6 Turbo Porter can be used in rough terrain and with a ceiling altitude of 10,000 metres even in high alpine terrain. On 18 August 2004, the Austrian Armed Forces received from the Ministry of the Interior its 13th Pilatus PC-6 Turbo Porter, which had previously flown as a calibrator for the Federal Office of Calibration and Surveying. The blue livery of this aircraft was retained after the takeover by the 4th Squadron of the 1st Air Regiment.

Agusta-Bell AB 212

Seite/page 106

Von 1975 bis 1980 stellte man beim Österreichischen Bundesheer Überlegungen betreffend einen Nachfolger für den Agusta-Bell 204B an. Obwohl bei einer Truppenerprobung beim Jägerbataillon 25 im September 1976 der bei Sud Aviation entwickelte Aérospatiale S.A.330 Puma wegen seiner größeren Transportkapazität besser abschnitt, wurden bei der Costruzioni Aeronautiche Giovanni Agusta SpA 24 AB 212 bestellt und zwischen März 1980 und April 1981 geliefert. Möglicherweise fiel diese Entscheidung deshalb zugunsten des AB 212 aus, weil er für die Piloten der scheidenden Agusta-Bell AB 204 keine große Umgewöhnung erforderte und ebenfalls von Pratt & Whitney PT-6-Turbinen angetrieben wird, wie die Pilatus PC-6 Turbo Porter und der Pilatus PC-7 Turbo Trainer.

From 1975 to 1980, the Austrian Armed Forces were considering a successor to the Agusta-Bell 204B. Although the Aérospatiale S.A.330 Puma, developed by Sud Aviation, performed better during a troop test at Jaeger Battalion 25 in September 1976 because of its greater transport capacity, 24 Agusta-Bell AB 212 were ordered at Costruzioni Aeronautiche Giovanni Agusta SpA and delivered between March 1980 and April 1981. Perhaps this decision in favour of the AB 212 was made because there was no great re-adjustment required for the pilots of the decommissioned Agusta Bell AB 204 and it was also powered by Pratt & Whitney PT-6 turbines, such as the Pilatus PC-6 Turbo Porter and the Pilatus PC-7 Turbo Trainer.

Seite/page 108

Die 24 Agusta-Bell AB 212 wurden auf die 1. Staffel des Fliegerregiments 1 in Langenlebarn und die 1. Staffel des Fliegerregiments 3 in Linz-Hörsching aufgeteilt. Der Hubschrauber wird für Lastentransporte bis zu 2300 Kilo Gewicht eingesetzt – auch mit einem 600 Liter fassenden Löschwasserbehälter, als Transporter für bis zu 13 Passagiere, als Sanitätshubschrauber mit sechs Krankentragen und zwei Sanitätern sowie als Rettungshubschrauber, ausgerüstet mit einer Seilwinde, zum Beispiel für Bergungen aus gebirgigem Gelände. Als Gefechtsfeldhubschrauber kann der AB 212 wahlweise eine leichte Flugabwehrkanone oder einen schweren Granatwerfer mit Geschützbedienung transportieren, aber ebenso eine Pioniergruppe mit Spreng- und Sperrmitteln.

The 24 Agusta-Bell AB 212 were split between the 1st Squadron of the 1st Air Regiment in Langenlebarn and the 1st Squadron of the 3rd Air Regiment in Linz-Hörsching. The helicopter is used for load transport up to 2,300 kilos (5,060 lb) – including a 600 litre fire water tank, as a transporter for up to 13 passengers, as a medical helicopter with six stretchers and two paramedics and as a rescue helicopter, equipped with a winch, for example, for salvage in mountainous terrain. As a battlefield helicopter, the AB 212 can either transport a light anti-aircraft gun or a heavy grenade launcher with gun operation, but also a pioneer group with explosive and blocking devices.

Im April 2010 wurde entschieden, die Agusta-Bell 212 des Österreichischen Bundesheeres bei der Agusta SpA in Cascina Costa einem „Midlife Upgrade“ unterziehen zu lassen, was eine Reihe von Modifikationen beinhaltet: zum Beispiel ein Head-up-Display, eine neue Funkausrüstung und passiven Selbstschutz, wie gepanzerte Pilotensitze und Kevlarmatten. Die Absicht dahinter ist, mindestens sechs AB 212 noch weitere 100.000 Flugstunden beziehungsweise 25 Jahre im Dienst zu halten.

In April 2010, it was decided to have the Agusta-Bell 212 of the Austrian Armed Forces undergo a “midlife upgrade” at Agusta SpA in Cascina Costa, which includes a number of modifications: for example, a head-up display, a new radio equipment and passive self-protection, such as armoured pilot seats and Kevlar mats. The intention behind this is to keep at least six AB 212s for another 100,000 flight hours or 25 years in service.

Pilatus PC-7 Turbo Trainer

Seite/page 110

Obwohl die Saab 91D noch im Betrieb stand, begann man beim Österreichischen Bundesheer Anfang der 1980er-Jahre, sich um einen Nachfolger der Safir umzusehen. Gegen die Konkurrenz der amerikanischen Beechcraft T-34 Turbo Mentor, der englischen NDN-1 Firecracker und der italienischen SIAI-Marchetti SF-260TP entschied man sich für 16 Pilatus PC-7 Turbo Trainer aus der Schweiz, von denen die ersten sechs Maschinen am 26. Jänner 1984 bei der Fliegerschule in Zeltweg ihren Dienst aufnahmen, womit das Österreichische Bundesheer nun zu den derzeit weltweit 21 Betreibern dieses Typs gehört.

Although the Saab 91D was still in operation, the Austrian Armed Forces began in the early 1980s to look for a successor to the Safir. Against the competition of the American Beechcraft T-34 Turbo Mentor, the English NDN-1 Firecracker and the Italian SIAI-Marchetti SF-260TP they decided on 16 Pilatus PC-7 Turbo Trainers from Switzerland, of which the first six machines began their service at the Flying School in Zeltweg on 26 January 1984, making the Austrian Armed Forces now one of the world's 21 operators of this type.

Seite/page 112

Nachdem ein Flugschüler in seiner Grundschulung ungefähr 140 Stunden auf der Pilatus PC-7 absolviert hat, folgt, abhängig von seiner Eignung und dem Bedarf, die Entscheidung, ob der angehende Bundesheerpilot zu den Hubschraubern oder den Flächenflugzeugen wechselt. Als künftiger Hubschrauberpilot wird seine Schulung auf der Sud Aviation S.A.316B Alouette III fortgesetzt oder als Pilot bei der leichten Transportstaffel auf Pilatus PC-6 Turbo Porter und – falls der junge Flieger zur „Speerspitze" der Österreichischen Luftstreitkräfte gehören möchte – auf der Saab 105OE, die ihm den Weg zum Eurofighter EF 2000 bereitet.

After completing approximately 140 hours of pilot training on the Pilatus PC-7 during basic training, the prospective Federal Armed Forces pilot will need to decide whether to switch to helicopters or fixed-wing aircraft, depending on their suitability and needs. As a future helicopter pilot, the training will be continued on the Sud Aviation SA316B Alouette III or as a pilot in the light transport squad on Pilatus PC-6 Turbo Porter and – if the young aviator wants to belong to the "spearhead" of the Austrian Air Force – on the Saab 105OE, which prepares his way for the Eurofighter EF 2000.

3H FD

Nach einer Umorganisation der Fliegerschule in Zeltweg im Jahr 1976 erlangte diese den Status einer Waffen- und Fachschule. Neben der Grund- und Fortgeschrittenenausbildung dient die Pilatus PC-7 dort zur Instrumenten-, Verbands- und Kunstflugschulung sowie zur taktischen Ausbildung. Zu diesem Zweck kann man den Turbo Trainer unter den Tragflächen mit zwei 12,7 mm Maschinengewehrbehältern HMP-250 der belgischen Waffenschmiede FN und zwei Behältern Thales LAU-32 mit je sieben ungelenkten 7 cm Raketen ausrüsten. Es können aber auch zwei Zusatztanks mit 145 oder 248 Liter Treibstoff mitgeführt werden.

After a reorganization of the aviation school in Zeltweg in 1976, this facility gained the status of a weapons and technical college. In addition to basic and advanced training, the Pilatus PC-7 is used for instrument-, formation- and aerobatic training as well as for tactical training. For this purpose, the Turbo Trainer can be equipped under the wings with two 12.7 mm machine gun containers HMP-250 of the Belgian armor FN and two containers Thales LAU-32 with seven unguided 7 cm rockets. But it can also carry two additional tanks with 145 or 248 litres of fuel.

Saab 35OE Draken

Seite/page 114

Da der Österreichische Staatsvertrag aber eine glaubwürdigere Luftraumüberwachung zur Sicherung der Neutralität erforderlich machte, sah sich das Österreichische Bundesheer im Jahr 1982 um einen aus Kostengründen gebrauchten, jedoch echten Abfangjäger um. Neben dem Saab J35D Draken standen British Aircraft Corporation (BAC) Lightning, Dassault Mirage 5 und Northrop F-5E Tiger II zur Auswahl. Gegen den inhärenten Widerstand von Teilen der politischen Opposition und mancher Medien entschloss sich die Bundesregierung am 2. April 1985 zum Ankauf von 24 allwettertauglichen Saab J35D Draken der schwedischen Flygvapnet.

However, since the Austrian State Treaty required more credible airspace surveillance to ensure neutrality, the Austrian Armed Forces were looking in 1982 for a genuine interceptor, second hand of course, for cost reasons. In addition to the Saab J35D Draken there were British Aircraft Corporation (BAC) Lightning, Dassault Mirage 5 and Northrop F-5E Tiger II to choose from. Against the inherent resistance of parts of the political opposition and some media, the Federal Government decided on 2 April 1985 to buy 24 all-weather Saab J35D Draken from the Swedish Flygvapnet.

Seite/page 116

Vor der Auslieferung nach Österreich wurden die bestellten Saab J35D modernisiert, erhielten das Glascockpit der Saab J35F, neue Avionik-Komponenten und wurden auf zusätzliche 1000 Flugstunden modifiziert. Aufgrund dieser Adaptierungen änderte man die Bezeichnung des Flugzeuges in Saab 35OE. Zwischen dem 6. Juni 1988 und dem folgenden Jahr wurden alle 24 Saab 35OE Draken nach Graz-Thalerhof überstellt und bildeten dort die 1. und 2. Staffel des Überwachungsgeschwaders.

Prior to delivery to Austria, the ordered Saab J35D were modernized, receiving the glass cockpit of the Saab J35F, new avionics components and were modified for an additional 1,000 flight hours. Because of these modifications it changed the name of the aircraft to Saab 35OE. Between 6 June 1988 and the following year, all 24 Saab 35OE Draken were transferred to Graz-Thalerhof and formed there the 1st and 2nd Flight of the Surveillance Squadron.

Die erste Bewährungsprobe für das Überwachungsgeschwader war der Bürgerkrieg in Jugoslawien im Sommer 1991, bei dem einige jugoslawische Flugzeuge in den österreichischen Luftraum eindrangen, was aber nach den regelmäßigen Patrouillenflügen der Saab 35OE Draken entlang der Staatsgrenze aufhörte. Damit war auch die permanente Kritik an den teuren Abfangjägern beendet.

The first test for the surveillance squadron was during the civil war in Yugoslavia in the summer of 1991, in which some Yugoslav aircraft invaded Austrian airspace, but this ceased after regular patrol flights of the Saab 35OE Draken along the borders. This ended the permanent criticism of the expensive interceptors.

Im Jahr 1994 nützte das Österreichische Bundesheer die positive Stimmung gegenüber dem Saab Draken im Land und beschaffte für diesen AIM-9P3 Sidewinder und vier Jahre später AIM-9P5 Luft-Luft-Kurzstreckenraketen. In den Jahren 1994 bis 1997 erhielten, wie zuvor die Saab Draken der Suomen Ilmavoimat (finnische Luftstreitkräfte), alle österreichischen Draken beim finnischen Flugzeughersteller Valmet Oy eine Kampfwertsteigerung und wurden damit zu Saab 35OE Mk.2 Draken.

In 1994, the Austrian Armed Forces used the positive mood towards the Saab Draken in the country and procured air-to-air short-range missiles for this AIM-9P3 Sidewinder and four years later AIM-9P5. In the years 1994 to 1997 all Austrian Draken received at the Finnish aircraft manufacturer Valmet Oy, as previously the Saab Draken Suomen Ilmavoimat (Finnish Air Force), a combat value increase and were thus named Saab 35OE Mk.2 Draken.

Am 25. November 2005 wurden die Saab 35OE Draken der Österreichischen Luftstreitkräfte mit einem Festakt in Zeltweg offiziell verabschiedet und die letzte Maschine am 31. Dezember desselben Jahres außer Dienst gestellt. Für die letzte Flugvorführung hatte eine Saab 35OE den schwarz-goldenen Sonderanstrich „Dragon Knights“ erhalten. Dieses Flugzeug und der anlässlich des Jubiläums „1000 Jahre Österreich“ rot-weiß bemalte Draken „Ostarrichi 996“ werden in der Militärluftfahrtsammlung in Zeltweg aufbewahrt.

On 25 November 2005, the Saab 35OE Draken of the Austrian Air Force had their official farewell ceremony at Zeltweg and decommissioned the last machine on 31 December of the same year. For the last flight display, a Saab 35OE had received the black and gold special painting “Dragon Knights”. This plane and the Draken “Ostarrichi 996” painted red-white on the occasion of the anniversary “1,000 years Austria” are kept in the military aviation collection in Zeltweg.

British Aerospace BAe 146

Als sich Ende der 1980er-Jahre ein Mangel an Transportraum bemerkbar machte, entschloss sich das Verteidigungsministerium, drei verschieden große Transportflugzeuge zu beschaffen. Dafür kamen die British Aerospace BAe 146, die Aeritalia G.222 und die CASA-Nurtanio CN-235 infrage. Die erste Bestellung betraf die BAe 146. Dieses Regionalverkehrsflugzeug erhielt im Oktober 1990 in England einen neuen Anstrich und ihre österreichische Zivilregistration. Kaum wurde dieses Vorhaben bekannt, begann die größte österreichische Tageszeitung eine Kampagne gegen die Geldverschwendung durch die Beschaffung dieser Flugzeuge. Dies führte letztendlich zur Aufgabe dieser Absicht. Die BAe 146 erreichte nie österreichischen Boden.

When in the late 1980s a lack of transport capacity was noticed the Ministry of Defence decided to procure three different sized transporter planes. The British Aerospace BAe 146, the Aeritalia G.222 and the CASA Nurtanio CN-235 were eligible. The first order concerned the BAe 146. This regional airliner received a new coat of paint and its Austrian civil registration in October 1990 in England. As soon as this project became known, the largest Austrian daily newspaper started a campaign against it alleging a waste of money by procuring this aircraft. This ultimately led to the abandonment of this intention. The BAe 146 never reached Austrian soil.

CASA-Nurtanio CN-235-300

Die CASA-Nurtanio CN-235 ist eine Gemeinschaftsentwicklung der spanischen Construcciones Aeronáuticas SA sowie der indonesischen PT Industri Pesawat Terbang Nurtanio und flog erstmals am 11. November 1983. Von 2000 bis 2002 wurde die CN-235-300 mit der spanischen Registrierung „EC-101“ vom Österreichischen Bundesheer zur Auslandsversorgung österreichischer Soldaten geleast, dafür erhielt dieses Transportflugzeug eine österreichische Militärkennung.

The CASA-Nurtanio CN-235 is a joint development of the Spanish Construcciones Aeronáuticas SA and the Indonesian PT Industri Pesawat Terbang Nurtanio and flew for the first time on 11 November 1983. From 2000 to 2002, the CN-235-300 with the Spanish registration “EC-101” was leased by the Austrian Armed Forces to supply Austrian soldiers abroad and therefore this transporter plane got an Austrian military registration.

Sikorsky S-70A Black Hawk

BUNDESHEER

Seite/page 122

Ende des letzten Jahrhunderts zeichnete sich beim Österreichischen Bundesheer ein Bedarf einer Ergänzung der in die Jahre gekommenen und weniger werdenden Agusta-Bell AB 212 ab. Zur Auswahl standen der von Aérospatiale, Agusta, Fokker und Messerschmitt-Bölkow-Blohm (MBB) entwickelte NH-90, der EH Industries EH-101, der Aérospatiale A.S.532 Cougar und der Sikorsky S-70A Black Hawk. Nach Evaluierung aller vier Typen entschied sich das Österreichische Bundesheer für den Black Hawk, der in vielen Bereichen einen deutlichen Vorsprung gegenüber seinen Konkurrenten hatte.

At the end of the last century, there was a need for the Austrian Armed Forces to supplement the aging and declining number of Agusta-Bell AB 212. The choices were the NH-90 developed by Aérospatiale, Agusta, Fokker and Messerschmitt-Bölkow-Blohm (MBB), the EH Industries EH-101, the Aérospatiale A.S.532 Cougar and the Sikorsky S-70A Black Hawk. After evaluating all four types, the Austrian Armed Forces decided in favour of the Black Hawk, which in many areas had a clear advantage over its competitors.

Seite/page 123

Am 6. Oktober 2000 trat einer der seltenen Fälle ein, dass das Verteidigungsministerium den Wünschen des Österreichischen Bundesheeres entsprach, sich ebenfalls für den Sikorsky S-70A-42 Black Hawk entschied und neun Maschinen bestellte, von denen die erste am 21. September 2002, zerlegt im Bauch eines Großraumtransporters Antonow An-124-100 „Ruslan“, in Österreich landete. Das offizielle „Fly in“ des Black Hawk erfolgte am 18. Oktober 2002 in Anwesenheit des Verteidigungsministers in Langenlebarn.

On 6 October 2000, one of the rare cases occurred that the Ministry of Defence followed the wishes of the Austrian Armed Forces, and opted for the Sikorsky S-70A-42 Black Hawk and ordered nine machines. The first arrived on 21 September 2002 in Austria, disassembled in the belly of a wide-bodied transporter Antonov An-124-100 "Ruslan". The official "Fly in" of the Black Hawk took place on 18 October 2002 in the presence of the Defence Minister in Langenlebarn.

Am 11. Dezember 2002 traf der letzte Sikorsky S-70A-42 Black Hawk in Österreich ein. Die neun Hubschrauber wurden der 1. Staffel des Fliegerregiments 1 in Langenlebarn zugeteilt, welche ihre Agusta-Bell AB 212 an Linz-Hörsching abgab. Neben Ausbildungsflügen, darunter auch Hochgebirgskurse für ausländische Militärpiloten, erfüllten die S-70A Aufgaben im Rahmen der Katastrophenhilfe und führten Einsätze im Ausland durch, zum Beispiel für die KFOR.

On 11 December 2002, the last Sikorsky S-70A-42 Black Hawk arrived in Austria. The nine helicopters were assigned to the 1st Squadron of the 1st Air Regiment in Langenlebarn, which delivered their Agusta-Bell AB 212 to Linz-Hörsching. In addition to training flights, including high mountain courses for foreign military pilots, the S-70A performed disaster relief and overseas missions, such as KFOR.

Lockheed C-130K Hercules

Seite/page 126

Bereits Mitte der 1960er-Jahre liebäugelten die österreichischen Luftstreitkräfte mit der Beschaffung von Lockheed C-130 Hercules, doch blieb dies aufgrund des meistens permanent niedrigen Budgets für die Landesverteidigung jahrzehntelang ein Wunschtraum. Da sich Anfang des neuen Jahrtausends aber das Dienstende der beiden Short SC.7 Skyvan, der einzigen größeren Transportflugzeuge des Österreichischen Bundesheeres, abzeichnete, wurde am 10. Mai 2002 mit Großbritannien ein Vertrag über den Ankauf von drei in den Jahren 2000 und 2001 eingelagerten Lockheed Hercules C Mk.1P der Royal Air Force unterzeichnet.

Already in the mid-1960s, the Austrian Air Force was toying with the acquisition of Lockheed C-130 Hercules, but this remained for decades, due to the mostly low budget for the National Defence, a pipe dream. However, at the beginning of the new millennium, it emerged that the life span of the two Short SC.7 Skyvan, the only larger transporter plane of the Austrian Armed Forces, was coming to an end, so a contract was signed on 10 May 2002 with the UK for the purchase of three Lockheed Hercules C Mk.1P of the Royal Air Force which were mothballed in 2000 and 2001.

Seite/page 128

Vor ihrer Auslieferung nach Österreich wurden die nun als Lockheed C-130K bezeichneten Transportflugzeuge, die sich von den amerikanischen C-130H nur durch ihre britische Avionik unterscheiden, bei der Marshall of Cambridge Aerospace Ltd. diversen Modifikationen (unter anderem eine neue Avionik, neues Radar und eine Luftbetankungssonde) unterzogen. Zwischen dem 23. März 2003 und dem 18. Februar 2004 nahmen die drei Hercules ihren Dienst bei der Lufttransportstaffel des Fliegerregiments 3 in Linz-Hörsching auf.

Prior to their delivery to Austria, the transport planes, now designated as Lockheed C-130K, which differs from the American C-130H only by their British avionics, were subjected to various modifications (including a new avionics, new radar and an aerial refuelling probe) by Marshall of Cambridge Aerospace Ltd. Between 23 March 2003 and 18 February 2004, the three Hercules took up duty at the air transport squadron of the 3rd Air Regiment in Linz-Hörsching.

AUSTRIAN AIR FORCE
8T CB

Seit dem Jahr 2004 fliegen diese drei Großtransportflugzeuge Personal und Nachschub zu den österreichischen Auslandskontingenten, zum Beispiel in den Kosovo und nach Syrien. Im Rahmen der Tsunami-Hilfe brachte eine C-130K Medikamente und Babynahrung nach Colombo, der Hauptstadt von Sri Lanka. Durch ein lufttransportfähiges MEDEVAC-Modul ist die österreichische Lockheed C-130K Hercules seit dem 8. April 2010 auch als fliegendes Spital einsetzbar, welches zwei Intensivpatienten oder neun Verwundeten sowie fünf Personen vom medizinischen Personal Platz bietet und bei der Rückführung verwundeter österreichischer Soldaten aus Bosnien und dem Kosovo bereits zum Einsatz kam.

Since 2004, these three large transporter planes have been flying staff and supplies to Austria's soldiers abroad, for example to Kosovo and Syria. As part of the tsunami relief, a C-130K brought medicines and baby food to Colombo, the capital of Sri Lanka. The Austrian Lockheed C-130K Hercules has also been deployed as a flying hospital since 8 April 2010 thanks to an air-transportable MEDEVAC module, which offers space for two intensive care patients or nine wounded and five persons from medical personnel and for the repatriation of wounded Austrian soldiers from Bosnia and Kosovo.

Northrop F-5E Tiger II

Seite/page 130

Anfang der 2000er-Jahre ergab sich für das Österreichische Bundesheer das Problem, dass der Flugbetrieb der seit 1988 im Dienst stehenden Saab 35OE Draken nur mehr mit immer größerem technischen Aufwand und zunehmenden Kosten aufrechtzuerhalten war. Gleichzeitig verzögerte sich die Auslieferung der für das Jahr 2002 erwarteten Eurofighter EF 2000, und ein Leihvertrag für sechs bis acht Maschinen dieses Typs auf drei Jahre scheiterte an den hohen Kosten von 300 Millionen Euro. Ein glücklicher Umstand für die österreichische Luftraumüberwachung war, dass die Schweiz in diesen Jahren von ihren Northrop F-5E Tiger II auf die McDonnell/Douglas F-18A/C Hornet umstieg und die Helvetier ihre F-5E verkauften. So wurde zwischen den Verteidigungsministern beider Länder eine Vereinbarung zur Anmietung von zwölf Northrop F-5E abgeschlossen, welche die zeitliche Lücke bis zum Eintreffen der Eurofighter in Österreich schließen sollten.

In the early 2000s, the Austrian Armed Forces faced the problem that the operation of the Saab 35OE Draken, which had been in service since 1988, could only be maintained with ever greater technical effort and increasing costs. At the same time, the delivery of the Eurofighter EF 2000 expected for 2002 was delayed, and a loan agreement for six to eight machines of this type for three years failed due to the high cost of 300 million euros. A happy circumstance for the Austrian air traffic control was that in these years Switzerland switched from its Northrop F-5E Tiger II to the McDonnell/Douglas F-18A/C Hornet and the Swiss sold their F-5E. An agreement was signed between the defence ministers of both countries to rent twelve Northrop F-5Es, which should close the time gap until the arrival of Eurofighters in Austria.

Seite/page 132

Nach der Zustimmung der US-Administration zu diesem Leihvertrag reisten am 14. Juni 2004 vier Piloten und 33 Mann Bodenpersonal des Österreichischen Bundesheeres zur Schulung auf die Northrop F-5E Tiger II nach Dübendorf in der Schweiz. Gegen Jahresende wurden weitere vier Draken-Piloten auf der F-5E geschult, und im folgenden Jahr waren es vier Saab-105OE-Piloten – bis schließlich 18 Piloten und 63 Techniker auf den Schweizer Jet eingeschult waren.

Following the US Administration's approval of this loan agreement, four pilots and 33 ground crew members of the Austrian Armed Forces travelled to the Northrop F-5E Tiger II to Dübendorf in Switzerland on 14 June 2004. At the end of the year, four more Draken pilots were trained on the F-5E, and the following year four Saab 105OE pilots – until finally 18 pilots and 63 technicians were trained on the Swiss jet.

SIEGE EJECTABLE
SCHLEUDERSITZ
SEGGIO EIETTABILE
DANGER!
GEFAHR!
PERICOLO!
2
2.D-Handgriff 2m ausziehen
1.Pousser le verrou
SAUVETAGE

Ab dem Juli 2005 nahmen die zwölf Northrop F-5E Tiger II den regulären Flugbetrieb über Österreich auf, wobei die Schweizer taktischen Nummern der Flugzeuge beibehalten wurden. Inzwischen hatten bereits einige Saab 35OE Draken aus Sicherheitsgründen Flugverbot erhalten, und die geleasten F-5E füllten die entstandene Lücke zur Zufriedenheit aller Beteiligten.

As of July 2005, the twelve Northrop F-5E Tiger II began regular flight operations over Austria, with the Swiss tactical numbers of the aircraft being maintained. Meanwhile, some Saab 35OE Draken had been banned for safety reasons, and the leased F-5E filled the gap to everyone's satisfaction.

Einen ihrer letzten Einsätze in Österreich flogen die Northrop F-5E Tiger II im Rahmen der Luftraumüberwachung während der Fußball-Europameisterschaft „Euro 08". Sie waren dabei ungefähr 70 Stunden in der Luft. Am 30. Juni 2008 startete letztmalig eine Rotte F-5E von Graz-Thalerhof aus zu einem Patrouillenflug. Zum danach erfolgenden „Fly out" waren vier Northrop F-5E Tiger II zum Schweizer Flugplatz Payerne gestartet und wurden dabei von zwei der im Juli 2007 in Dienst gestellten Eurofighter EF 2000 des Österreichischen Bundesheeres bis zur Staatsgrenze begleitet.

One of their last missions in Austria flew the Northrop F-5E Tiger II in the context of airspace monitoring during the European Football Championship "Euro 08". They were in the air for about 70 hours. On 30 June 2008, a pack of F-5E started for the last time from Graz-Thalerhof for a patrol flight. Four Northrop F-5E Tiger II had departed for the Swiss airfield Payerne and were accompanied to the state border, by two of the Eurofighter EF 2000 of the Austrian Armed Forces, which was put into service in July 2007.

Eurofighter EF 2000

NGAR II

Seite/page 135

Ende des Jahres 2000 richtete Österreich eine „Request for information“ an die Hersteller der Boeing F/A-18C/D Super Hornet, Dassault Mirage 2000-5, Lockheed-Martin F-16C/D Block 50 und der Saab JAS-39 Gripen. Die Exportfirma RAC aus Russland hatte bereits im September davor ein Paket von Mikojan-Gurewitsch MiG-29SMT und UBT (Zweisitzer) angeboten. EADS wurde nicht gefragt, reklamierte sich aber nachträglich in das Verfahren.

At the end of the year 2000, Austria sent a “Request for Information” to the manufacturers of the Boeing F/A-18C/D Super Hornet, Dassault Mirage 2000-5, Lockheed-Martin F-16C/D Block 50 and the Saab JAS-39 Gripen, The export company RAC from Russia had already offered in September before a package of Mikoyan-Gurevich MiG-29SMT and UBT (two-seater). EADS was not asked, but later complained.

Seite/page 136

Ende der 1990er-Jahre war ein Großteil der Angehörigen der österreichischen Luftstreitkräfte der Meinung, dass als logischer Nachfolger für den Saab 35OE Draken der Saab JAS-39 Gripen beschafft werden solle. Es war angedacht, in Zeltweg ein Technikzentrum für den Gripen zu errichten und dort sowohl die eigenen als auch jene JAS-39 ausländischer Luftwaffen zu betreuen, denen der Weg nach Schweden zu weit ist. Allerdings besitzt der Gripen eine für seine Aufgaben in Österreich zwar ausreichende, aber doch um einiges geringere Leistungsfähigkeit als der Eurofighter, war aber nur um wenige Prozent preiswerter als dieser. Deshalb traf die österreichische Regierung am 2. Juli 2002 die Entscheidung, 24 einsitzige Eurofighter EF 2000 zu bestellen.

At the end of the 1990s, a large part of the members of the Austrian Air Force was of the opinion that the Saab JAS-39 Gripen should be acquired as the logical successor to the Saab 35OE Draken. It was planned to build a technical centre for the Gripen in Zeltweg and to look after both their own and those JAS-39 of foreign air forces, who were geographically too far from Sweden. However, the Gripen, sufficient for Austria's needs, but still with a much lower performance than the Eurofighter, was only a few percent cheaper than this. Therefore, on 2 July 2002, the Austrian Government decided to order 24 single-seater Eurofighter EF 2000s.

Die Beschaffungskosten des neuen Abfangjägers lagen über der politischen Schmerzgrenze. Dem damaligen Verteidigungsminister gelang es, die Anzahl der bestellten Eurofighter EF 2000 von 24 auf 18 zu reduzieren. Dadurch sah sich das Österreichische Bundesheer plötzlich dazu gezwungen, sein Luftraumüberwachungskonzept an die verringerte Anzahl seiner Eurofighter anzupassen. Von den erwarteten sechs zweisitzigen EF 2000 zur Schulung war keine Rede mehr.

Since the procurement costs of the new interceptor were above the political pain threshold, the then Minister of Defence managed to reduce the number of Eurofighters EF 2000 ordered from 24 to 18. Therefore, the Austrian Armed Forces were suddenly forced to adapt its airspace surveillance concept to the reduced number of its Eurofighters. The six two-seat EF 2000 for training purpose was mentioned no more.

Austrian
7L WB
7L WF
NO STEP
DANGER

Seite/page 139

Während des Wahlkampfes zur Nationalratswahl Ende September 2006 versprachen die österreichischen Sozialdemokraten den Wählern, im Falle der Übernahme der Regierung aus dem teuren Eurofighter-Kaufvertrag auszusteigen. Nachdem sie die Wahl gewonnen hatten, fand sich allerdings kein rechtlicher Grund, dieses Versprechen umzusetzen. Der neue Verteidigungsminister konnte jedoch eine weitere Reduktion der bestellten Eurofighter EF 2000 auf nunmehr 15 Exemplare erreichen. Auch vorgesehene Teile der technischen Ausrüstung wurden nun nicht mehr beschafft, sodass es sich bei dem österreichischen Eurofighter nur mehr um eine abgespeckte Version handelt. Die nun auch schnellere Abnutzung von nur 15 Eurofighter EF 2000 in einem Aufgabenbereich, der für 24 Maschinen vorgesehen war, war für die österreichische Politik bisher kein Thema.

During the election campaign for the Parliamentary Elections at the end of September 2006, the Austrian Social Democrats promised voters to step away from the expensive Eurofighter purchase agreement in the event of them taking over government. After winning the election, however, there was no legal reason to enforce this promise. However, the new Minister of Defence was able to achieve a further reduction of the ordered Eurofighter EF 2000 to 15 units. Also the planned parts of the technical equipment were no longer bought, so this Austrian Eurofighter is only a stripped-down version. The now even faster wear and tear of only 15 Eurofighter EF 2000 in an operational area intended for 24 machines was so far no issue for the Austrian politicians.

Seite/page 140

Am 12. Juli 2007 landete der erste österreichische Eurofighter EF 2000 in Zeltweg, am 24. September 2009 traf die letzte Maschine dort ein. Ihre Aufgaben im Rahmen der Luftraumüberwachung übernahmen die Eurofighter bereits davor, am 1. Juli 2008. Die volle Einsatzbereitschaft aller EF 2000 wurde 2015 erreicht. Die Luftraumüberwachungs-Flugzeuge fliegen bis zu 1100 Stunden pro Jahr, wobei sich die Eurofighter ihren Überwachungsauftrag mit dem letzten Dutzend einsatzfähiger Saab 105OE der Düsentrainerstaffel teilen.

On 12 July 2007, the first Austrian Eurofighter EF 2000 landed in Zeltweg, on 24 September 2009, the last machine arrived there. Eurofighter took over their tasks in the context of airspace surveillance already before that, namely on 1 July 2008. The full operational readiness of all EF 2000 was achieved in 2015. Airborne surveillance aircraft fly up to 1,100 hours a year, with Eurofighters sharing their surveillance mission with the last dozen deployable Saab 105OE of the jet training squadron.

Diamond DA40 NG

Zwischen dem 3. Mai und dem September 2018 erhielten die österreichischen Luftstreitkräfte die ersten beiden von insgesamt vier Diamond DA40 NG zur Pilotengrundausbildung.

Between 3 May and September 2018, the Austrian Air Force received the first two of a total of four Diamond DA40 NG for basic pilot training.

Bildnachweis / Picture credits

BMLV Gerhard Simader: 109
Fliegerhorstbataillon 2, Bildstelle: 66
Wernfried Haberfellner: 13
Heeresbild- und Filmstelle: 10, 11, 12, 13, 14, 15, 16, 17, 18, 19, 20, 21, 22, 23, 24, 25, 27, 30, 31, 33, 34, 38, 39, 41, 43, 44, 45, 48, 49, 50, 54, 55, 57, 58, 59, 62, 63, 64, 65, 68, 69, 70, 79, 93, 94, 96, 97, 98, 99
Wilhelm Hesz: 7, 36
Rudolf Höfling: 29, 32, 35, 47, 60, 71, 72, 73, 74, 75, 77, 82, 83, 88, 89, 90, 100, 103, 104, 108, 116, 118
ÖBH: 76, 85, 101, 102, 106, 110, 111, 113, 114, 117, 120, 121, 125, 128, 129, 130, 132, 133, 135, 136, 138, 140
Gerald Penz: 26
Offzstv. Thomas Rotschek: 78, 86, 87
Hubert Strimitzer: 52, 53
Erich Strobl: 37, 40, 61, 80, 81, 84, 105, 119, 122, 123, 126, 134, 139, 142
Paul Vaszarics: 28, 42, 46, 51, 56, 67

Quellenverzeichnis / Bibliography

Hainzl, Wolfgang; Hauke, Erwin: Die Fliegerkräfte Österreichs 1955 bis heute. Graz 1987.
ÖFH Nachrichten: Vereinsmagazin der Österreichischen Flugzeug Historiker. Verschiedene Ausgaben.
Strobl, Erich: Österreichische Luftstreitkräfte – Austrian Air Arms. Luftwaffen Profile 2. 2012.

Danksagung / Acknowledgments

Der Autor bedankt sich bei Georg Mader und Erich Strobl für ergänzende Informationen.
The author wants to thank Georg Mader and Erich Strobl for additional information.